THEORY OF EVERYTHİNG

MUSTAFA KILIÇ

Published by MUSTAFA KILIÇ, 2025.

THEORY OF EVERYTHİNG

First edition. June 12, 2025.

ISBN: 979-8231458363

Written by MUSTAFA KILIÇ.

Table of Contents

Published by Mustafa Kılıç
First Edition, 2025

THEORY OF EVERYTHING

MUSTAFA KILIÇ

PREFACE

This is my seventh attempt at rewriting this book from scratch... I hope this time it's the final one!

Condensing a theory that explains all of nature into a single book is a formidable challenge. With countless natural phenomena to choose from, I struggled to decide what to include. During the writing process, I either uncovered more critical topics or grew dissatisfied with my narrative style, prompting me to start over. For example, when discussing photons within the theory, I must explain colors—a topic so vast that my photon discussion stretches on endlessly. The theory of everything is so expansive that it could not fit into one book, nor even thousands of encyclopedic volumes. Still, I hope this seventh attempt achieves the success I envision.

The purpose of this book is to elucidate the quantum mechanics of atoms and larger particles within the framework of the theory of everything. Once you grasp how quantum mechanics operates, you can apply the laws I've outlined to unravel the mysteries of any natural phenomenon not covered in this book—in mere seconds, just as I do. This theory is systematic, and its system is broader than the theory itself. By integrating this theory with your existing knowledge, you can use its system to generate countless new insights consistent with nature. In other words, by mastering this theory through this book, you will gain far more than its pages contain.

Throughout the book, I will repeat certain quantum elements and behaviors to ensure they stick in your mind, as I have redefined some concepts. Starting from before the Big Bang and progressing to the present, we will unravel the mysteries of quantum mechanics and the universe step by step.

A PHILOSOPHER FROM HUMBLE BEGINNINGS

In 1976, I was born in a small town near Turkey's western coast, the first child of a modest family. I vaguely recall my mother teaching me to read before I even started school. My childhood lacked internet, television, smartphones, or computers. Reading must have captivated me early on, as I devoured anything with text—newspapers, books, flyers, or even roadside signs. When I began elementary school, I discovered the library. During holidays, I'd immerse myself there, reading every book without discrimination. Despite never doing homework, I breezed through to the seventh grade. But at age ten, I stumbled upon comic books and read nothing else. That summer, I started working and earned my first money, which felt like a fortune. Foolishly, I decided, "Why bother with school when I can earn this much as a kid?" and dropped out.

My life became a cycle of unemployment, idleness, and temporary jobs. In 2009, with no capital, I launched my own business. It grew with loans, but in 2015, I went bankrupt. For days, I racked my brain for solutions, but found none. Then I thought to myself, "If I spent as much time pondering the universe's mysteries as I did trying to save my business, I'd have solved them all." I wasn't married, nor did I have a spouse or children to care for. So, I decided to become a philosopher. After all, anyone who declares themselves a philosopher can be one, right? I first devised a unique business idea, then conceptualized an unparalleled universal constitution and economic system. Perhaps I'll write those down someday. Next, I asked myself, "As a philosopher, what's next? What should I solve?" A question struck me:

“Why is there everything instead of nothing?”

Why does everything exist? Why do I exist? Where did the microscopic particles that form my surroundings come from, how did they form, and how did they combine to create such a perfect environment? Until this question arose, my idea of philosophy was entirely different. Finding an answer was immensely challenging. Beyond seeking answers, contemplating nothingness within everything or infinity within my finite surroundings drove me to the brink of madness. Back then, my physics knowledge was limited to a few science documentaries and news articles—often unverified, incomplete, or misleading. Most of the physics in this book I learned later, through developing my theory.

Imagining infinite nothingness in three dimensions and trying to discern what existed first was mentally exhausting. I could physically feel it, as if my brain's blood vessels were swelling like the jugular in my neck. For days, I paced among machines, stepped outside the shop, or sat straining my mind. Eventually, I stopped trying to imagine infinity. Whenever I pictured a boundary or end to infinity, I was merely thinking of a finite set. Don't try to imagine infinity! Our brains work algorithmically, and even when we try to conceive of infinity, we instinctively impose an end. After days of reflection, I discovered what existed first in infinite nothingness: everything that exists now was there, but in a singular state. (You'll understand this better later.) In that moment, a light sparked in my dark mind, like a lightbulb flashing in a cartoon. This revelation eased all my mental strain, and I still sometimes feel that moment vividly.

INTRODUCTION TO THE THEORY OF EVERYTHING

Modern scientific theories are built on the deductive method. However, no theory has reached back beyond the Big Bang to the point of "nothingness." Consequently, existing theories start with observed phenomena and aim for specifics, but they fail to arrive at the origin of everything. This may be why a "theory of everything" has yet to be found. Induction and deduction complement each other. Unknowingly, I began with deduction, using it solely as a reference point: the "nothingness" in the question, "Why is there everything instead of nothing?" Infinite nothingness enabled me to discover the theory of everything.

After identifying the first natural event in infinite nothingness, I made thousands of discoveries, building the theory step by step from the specific to the present. The theory of everything, advancing inductively from nothingness, was completed without error. I claim its flawlessness because it relies on a single fundamental force, which allowed no deviation. When I strayed onto a wrong path during induction, the theory produced errors. Recognizing these, I worked to find the correct course. Because the theory progresses inductively, anyone—you, someone from a thousand years ago, or a person born a century from now—could reach the same conclusions.

Once you grasp this theory, you can unravel the micro and macro mechanics of any unfamiliar natural phenomenon in seconds. You don't need to be a physicist or scientist to understand it. The theory of everything explains all phenomena using a single fundamental force, one everyone knows and uses in daily life. Before delving into the theory, I want to review or introduce key concepts: What is knowledge? What are physical and metaphysical knowledge? How do we distinguish physics from metaphysics? What is a theory, and what makes a good one? I will provide existing answers to these questions, enhanced with my own insights.

WHAT IS KNOWLEDGE?

Knowledge encompasses everything—physical and metaphysical—that the human mind and logic can grasp. Simply put, anything you understand is knowledge to you.

All knowledge produced by humanity, now and in the future, consists fundamentally of definitions and interpretations. We define and name our discoveries and inventions based on their characteristics, thereby creating knowledge. We also assign names to the dynamic events involving physical elements. These events are defined by their properties and then interpreted. Thus, we must acknowledge that all physical knowledge is rooted in definitions and interpretations. The same event or element can be interpreted differently—for instance, Flat Earth versus Geoid Earth, or gravity versus celestial thrust.

I categorize knowledge into two types: physical and metaphysical. Each is further divided into three subtypes: 1) True knowledge, 2) False knowledge, and 3) Unfalsifiable knowledge. Determining whether knowledge is true, false, or unfalsifiable depends on an individual's intelligence.

METAPHYSICS AND PHYSICS

I propose that everything exists. Everything in our minds exists, and the physical foundation of all that exists resides as knowledge in our brains or in nature. This knowledge can be classified as true, false, or unfalsifiable.

I aim to support this proposition with a chart.

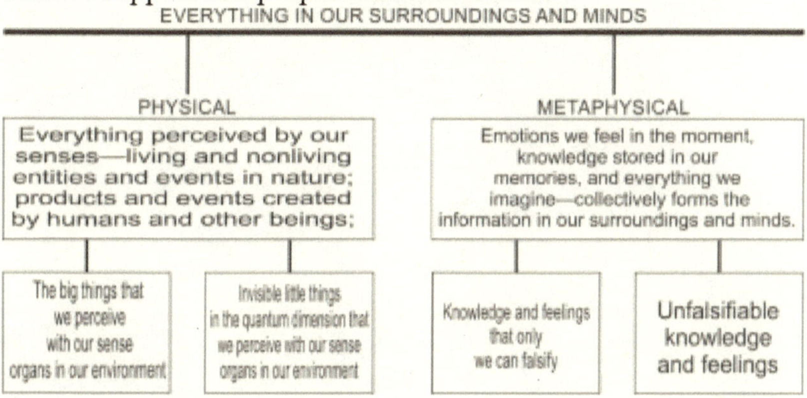

Having defined physics and metaphysics, I categorized each into two groups. Now, I'd like to provide more detail about these physical and metaphysical entities.

Physical entities are composed of indivisible, smallest, singular micro-particles—atoms. According to the laws of thermodynamics, these atoms can neither come into existence from nothing nor cease to exist. Atoms can separate from or combine with other atoms based on environmental factors. During this process, large natural entities like stars, planets, mountains, and seas can "come into being" randomly and naturally, in name and knowledge. Similarly, human-made physical objects such as cars, planes, and phones are brought into existence by name. Here, "coming into being" means acquiring a defined or definable name. For example, flour and water have distinct definitions; when combined, they form dough. This set of elements, fundamentally made of atoms, becomes "dough" by name—a process of coming into being. This is precisely why all knowledge is metaphysical.

The laws of thermodynamics state that what cannot come into being from nothing or cease to exist are the smallest singular particles—atoms—that form physical entities. Moreover, these singular particles are indivisible. (I will explain why they cannot be divided later.) The characteristics distinguishing physical and metaphysical entities are as follows:

METAPHYSICS AND PHYSICS

Physics: Physical entities are composed of micro-particles. These particles can combine to form entities that "come into being" by name, or separate, causing entities to cease to exist both in name and form. Physical entities and events are perceivable by living beings and devices. The micro-particles forming these entities have existed since the Big Bang, independent of humans, and will exist eternally.

Metaphysics: Everything in our minds—every piece of knowledge—is metaphysical. Only what we perceive in our surroundings at a given moment is physical. For instance, imagine you're at home; everything you perceive around you is physical. When you leave for work, those things become metaphysical in your mind. Simply put, metaphysical entities persist relative to the algorithmic changes in our brains. Thus, metaphysical knowledge and thoughts can violate the laws of thermodynamics based on these changes. We can mentally create entities and events from nothing or make them disappear. When we cannot perceive their physical effects, everything is metaphysical. For example, you see a tree in the distance—it's physical. When you turn away, the tree becomes metaphysical knowledge to you. When you look again, it's physical. If a friend beside you sees the tree behind you and describes it, the tree is physical to them but metaphysical to you.

As shown in the diagram, I've divided physics and metaphysics into two groups each:

Physics Group 1: Large entities we perceive with our senses—houses, cars, other living beings, forests, seas, rivers, water, soil—elements and events we can see, hear, smell, touch, or taste.

Physics Group 2: Effects caused by quantum-scale particles perceived through our senses or devices—scattered atomic groups forming air, viruses, electricity, light, dark matter, bacteria, and more. Elements too small or distant to see directly emerge as knowledge through scientific theories, consisting of definitions and interpretations. Such knowledge has evolved, developed, or vanished over time.

Metaphysics: Encompassing all feelings and knowledge, metaphysics is a vast field. It includes both physical and non-physical knowledge. Abstract metaphysical concepts can be made concrete by artists (e.g., paintings, sculptures, films). Similarly, concrete physical entities and events can be abstracted in our minds. For example, if you want to modify your car, you can mentally disassemble its parts, add new ones, and imagine its appearance. Then, by acquiring the physical parts, you can turn your imagined car into reality.

Metaphysics Group 1: Falsifiable knowledge and feelings—essentially, all physical and scientific knowledge.

Metaphysics Group 2: Unfalsifiable knowledge and feelings known only to you or another—beliefs, love, hate, pain, suffering, parallel universes, time, paradoxes, and more.

THE LAWS OF THERMODYNAMICS

Before delving into the theory of everything, I'd like to introduce the laws of thermodynamics for those unfamiliar with them. These laws are remarkably robust; no experiment or natural phenomenon has ever disproven them. If a theory violates the laws of thermodynamics, I recommend discarding it immediately or viewing it as a metaphysical tale. Any theory that contradicts these laws is flawed or incomplete. In the theory I present, there will be no propositions that violate thermodynamics; in fact, I aim to conceptually strengthen these laws further.

The laws of thermodynamics are fundamental principles that define energy and heat transfer processes. There are four primary laws of thermodynamics. Let's explore them in detail:

(i) Zeroth Law:

If two systems are in thermal equilibrium with a third system, they are also in thermal equilibrium with each other.

Details: This law defines the concept of temperature and enables its use as a measurable quantity. It also reflects the principle of causality. The principle of causality asserts that every event or state has a cause, and every cause produces an effect. It states that events in the universe are not random but occur within a specific cause-and-effect framework, meaning an event requires a triggering cause to take place.

Key Features of the Principle of Causality:

- Specific Cause: Every effect has a cause.
- Temporal Order: The cause precedes the effect.
- Reproducibility: Identical causes under identical conditions yield identical effects.

  The principle of causality underpins scientific research and plays a crucial role in understanding physical laws. It enables scientists to identify cause-and-effect relationships through observations and experiments, explaining natural phenomena and predicting future events.

  In the theory of everything, I will adhere to the principle of causality. The physical behaviors of matter are not variable; the way matter interacts and reacts today is the same as it did in the past and will in the future.

(ii) First Law (Conservation of Energy):

Energy can neither be created nor destroyed; it can only transform from one form to another.

Details: The change in a system's internal energy equals the heat added to the system minus the work done by the system. Energy is a fundamental concept representing the capacity of physical systems to do work. The force generating energy arises from physical changes in a substance's position, structure, or surrounding environment in a chain reaction. The physical existence of matter is primary, while energy is secondary. Thus, matter can neither be created nor destroyed. Here, by "matter," I refer to the smallest indivisible unit—the atom. By examining the atom model in the theory of everything and observing its behavior in nature, you will gain a clearer and more logical understanding of atoms.

(iii) Second Law (Entropy):

In an isolated system, entropy always increases or at the very least remains constant; a completely ordered system becomes disordered over time.

Details: Natural processes tend toward energy dispersion and an increase in disorder (entropy). Heat naturally flows from a warmer region to a cooler one, not the reverse. When a cluster of numerous micro-particles begins to separate, irreversible changes occur. I will explain the quantum-scale functioning of entropy and why it is irreversible later.

(iv) Third Law:

As a system approaches absolute zero temperature (0 Kelvin), its entropy approaches zero.

Details: When a system nears absolute zero, it assumes a perfect crystalline structure, and its entropy minimizes. Absolute zero implies zero motion or change at the quantum scale—essentially, stillness. However, everything exhibits micro-scale vibrations (minimal positional changes). Since every group of matter is surrounded by other elements and everything interacts, a vibration-free or change-free state is impossible. In nature, absolute zero—or complete stillness—is unattainable. At the quantum level, matter lacks concepts of hot and cold; I will explain why later.

I have elaborated on the laws of thermodynamics with added details. Record these enhancements and ensure that the theory of everything does not violate these laws.

WHAT IS A THEORY? WHAT MAKES A GOOD THEORY?

A theory is a systematically organized set of rules and laws that explains numerous phenomena and forms the foundation of a scientific discipline—in other words, a theoretical framework.

You may be familiar with physical laws or have heard of them. Physical laws are propositions with inherent certainty, formed by interpreting and defining natural phenomena in a way that aligns with a theory. For example, consider two theorists, Ali and Veli, who developed distinct theories. While chatting over tea in an apple orchard, they observed an apple falling from a branch. Ali attributed the event to the earth pulling the apple, aligning it with his theory, and declared gravity a physical law. Veli, interpreting the same event through his theory, argued, "The sky pushed the apple down," and proclaimed celestial thrust a physical law. Thus, the same natural event led to two opposing physical laws. Since physical laws carry certainty within a theory, theorists rarely accept their definitional violation, as doing so would undermine their theory's foundation, rendering it invalid. Yet, regardless of interpretations, the apple's fall is a fact—nature continues to function. Although physical laws may differ in interpretation and definition, what matters is that they do not contradict natural phenomena or experiments. Both Ali and Veli's laws explain the apple's fall, making them valid. Therefore, a theory's strength depends on its physical laws aligning with natural phenomena. If a theory's laws are consistent with natural events, it is a correct and robust theory.

In short, a theory is a set of rules and laws constructed by a theorist—by which I mean the systematic arrangement of real natural phenomena into coherent sentences. As noted in dictionaries, it must be systematically organized and capable of explaining numerous phenomena. The more natural phenomena a theory can explain, the better it is.

Today, every scientific discipline is grounded in a theory, but none of these theories are systematically consistent with all of nature. When we consider nature as a whole—both visible and invisible phenomena—no theory explains everything, making current theories incorrect in the broader context of nature. For instance, at the macro level, Albert Einstein's relativity theories may seem correct; at the micro level, quantum field or particle theories may hold true. However, these theories conflict with each other, losing their systematic coherence. This is why the theory of everything will usher in a new era in science and create a significant stir.

Currently, each scientific discipline relies on a theory, but these disciplines often adopt independent theories. Unfortunately, accepted science consists of theories that are relatively false, incomplete, or contradictory to each other, yet progress within their own domains. If a discipline's foundational theory is fundamentally flawed, it builds error upon error. As the system legitimizes these errors, correcting them becomes increasingly difficult. Until the theory of everything is discovered, science has had to advance by compounding errors. Once the theory of everything emerges, a new era in science—from biology to quantum physics, astrophysics to chemistry—will begin. Upon finishing this book, you will decide whether the theory of everything has truly been discovered.

WHAT MAKES A GOOD THEORY?

A good theory advances scientific knowledge, explains natural phenomena, and possesses predictive power. The essential characteristics of a good theory are as follows:

1. Testability:

A theory must be testable through experiments and observations. It should generate hypotheses that can be experimentally verified or disproven.

In this book, while explaining the theory of everything, I will present numerous tested or testable hypotheses, along with dozens of analogies to illustrate concepts.

2. Falsifiability:

A theory must be falsifiable, meaning it should allow for situations or experiments that could potentially disprove it. As Karl Popper emphasized in his philosophy of science, a theory must be falsifiable to be considered scientific.

The theory of everything consists entirely of falsifiable propositions or those that can be supported by analogies.

3. Consistency:

A theory must be internally consistent and free of contradictions.

The theory of everything, relying on a single fundamental force, progresses systematically through induction, ensuring a high degree of consistency.

4. Simplicity (Ockham's Razor):

A theory should avoid unnecessary complexity and provide the simplest possible explanation. When multiple theories explain the same phenomenon, the simplest one should be preferred.

The theory of everything is remarkably simple—so much so that I might claim it to be the simplest theory. Upon reading this book, even a non-physicist adult will easily grasp it, while expert physicists may be astonished by its simplicity. If this theory is one day turned into an animated documentary, even children will understand it.

5. General Applicability:

A theory should explain a wide range of phenomena and be applicable to various situations, addressing not only specific events but also general principles.

The theory of everything effortlessly explains all observed and experimentally verified physical phenomena. While not every explanation can fit in this book, once you learn nature's operating principles, you'll be able to unravel any natural mystery in seconds.

6. Predictive Power:

A theory must accurately predict unobserved events or conditions, and these predictions should be testable and verifiable.

The theory of everything can generate countless hypotheses, some of which I'll include in this book. In the future, new observational discoveries can be easily explained using this theory.

7. Empirical Support:

A theory must be supported by observational and experimental data, backed by a broad base of evidence that validates and strengthens it.

You will witness that the theory of everything is supported by naturally occurring events and existing experiments.

8. Adaptability:

A theory should be revisable or expandable with new data and discoveries. As science progresses, it must remain open to development and updates.

The theory of everything is open to growth and advancing science. Once accepted, millions of new articles will be waiting to join the knowledge pool.

9. Explanatory Power:

A theory should provide satisfying answers not only to “how” but also to “why,” explaining complex events and processes in an understandable way.

These characteristics enhance a theory’s scientific validity, reliability, and utility, contributing to the advancement of scientific knowledge.

Additional Characteristics:

1. Non-Contradiction:

The laws and rules within a theory must not contradict each other. Paradoxes, or contradictions, may exist in metaphysics but not in nature or physics, as no contradiction has ever been observed or recorded in nature. The double-slit experiment, often cited as paradoxical, has simply been misinterpreted—I will offer my interpretation in this book. A nature containing paradoxes could neither function nor sustain life. In other words, for a theory to be systematic, it must not include laws that violate the principle of causality.

2. Comprehensiveness:

A theory should explain a wide range of experiments and natural phenomena. Since natural phenomena already occur, a theory need only provide logical and rational explanations for them. For experiments, existing ones should be explained, and hypotheses for new experiments should be generated. The more hypotheses a theory can produce, the better it is.

3. Falsifiability (Reiterated):

A good theory must be falsifiable. One of the key requirements in natural science is to offer observable and perceivable propositions. Such propositions can be tested, allowing their truth or falsehood to be verified. If a proposition or hypothesis cannot be tested, it should at least be supported by an analogy.

Can a Theory Without Mathematical Operations Be a Good Theory?

This depends on the propositions within the theory. Mathematics is a human invention—a constructed language, much like English, Spanish, or Turkish. In the language of mathematics, numbers, equations, and a few words are used. Humans have observed the elements and events of nature, defining each based on its properties. For instance, a stone with three sides was named a "triangle," a circle drawn on the ground was called a "circle." Counting emerged while gathering fruits into baskets, and addition, subtraction, division, and multiplication were invented during trade. Over time, this process led to the development of modern mathematics. For thousands of years, humans have added new inventions to mathematics, describing natural phenomena with equations and creating mathematical symbols for everything. Consequently, if a theory is falsifiable, a skilled physicist can express its hypotheses and propositions using mathematical operations and equations. I believe the theory of everything will lead to groundbreaking new equations in mathematics and solve some unresolved mathematical problems. A talented mathematician could uncover the hidden treasures in this book and present them to the world in mathematical language.

BEFORE DIVING INTO THE THEORY OF EVERYTHING

Before delving into the theory of everything, I'd like to define the three classes into which I've categorized everything in space at the micro/quantum level.

Everything in space is composed of atoms. The word "atom" derives from the ancient Greek term atomos, meaning "indivisible." In Turkish, I've defined the atom as an acronym: Ayrışmaz/Bölünmez/Parçalanmaz Tekil Organik Madde (Indivisible/Singular Organic Matter, or ATOM). However, since the acronym differs in English, you can simply note that atoms are indivisible and continue reading. Everything in nature is made of atoms, including light. Atoms exist in two primary states: solid and soft. When solid, small-volume atoms are vacuumed by other atoms, they stretch like rubber, becoming soft, flexible, and larger in volume. In their soft form, atoms form the foundational structure of light, depending on the number in the cluster.

Instead of assigning new names to the elements I've categorized into three classes, I've redefined existing terms in physics based on their properties. The scientific community can rename them if needed in the future.

THE THREE MAIN CLASSES IN SPACE

In the theory of everything, I've divided everything in nature—each a cluster of atoms—into three main classes: dark matter, elements, and photons/electrons.

1. Dark Matter

In the theory of everything, dark matter refers to clusters of an unknown number of symmetrical, solid atoms that form space. These clusters take on a spherical shape as atoms aggregate. As the number of dark matter clusters increases, external-to-internal pressure and the internal-to-external pressure they exert on each other cause them to adopt symmetrical forms. When a dark matter cluster breaks apart, its symmetry is disrupted, resulting in smaller, asymmetrical clusters. The atoms on the surface of symmetrical dark matter clusters do not wear down due to interactions (e.g., friction), but their roots remain solidly attached to the cluster, allowing them to stretch into a soft state from the surface outward. Thus, the atom's root is solid, while its tip becomes soft. (In this book, dark matter and dark energy are the same; their names only vary based on location and surroundings.)

2. Elements

Elements are asymmetrical clusters of solid atoms that emerge when some dark matter clusters break apart. Since the atom model in the theory of everything differs from the conventional one, I suggest setting aside the current model—including protons, neutrons, bosons, and similar components—until you've read the book.

3. Photons and Electrons

Photons are clusters formed when solid atoms from disintegrating dark matter or from the subsequent separation, dispersion, or decay of elements transition into a soft state. In the theory of everything, clusters of soft atoms are called "photons" when in motion, i.e., in the form of light. When these soft atoms are positioned close to their source element or another element, I've redefined their form as "electrons."

As you've read, elements emerge from some dark matter clusters, and photons and electrons arise from elements.

The theory of everything, using a single fundamental force and a single fundamental structure—atoms—easily explains every natural phenomenon and experimental event through defined components. I will describe how space, the Big Bang, the universe, star systems, black holes, viruses, living beings, earthquakes, and countless other phenomena form and occur. Once you grasp the theory, you and scientists will be able to unravel nature's mysteries.

Let's begin our journey from absolute nothingness, from the void...

INTRODUCTION TO THE THEORY OF EVERYTHING

Understanding the workings of the universe has been one of humanity's greatest and most enduring passions. Since ancient times, philosophers, scientists, and thinkers have sought to uncover the fundamental principles of the universe, developing numerous theories along the way. These theories have opened many doors to explaining nature's mysteries, yet each time, they have led to even larger questions.

Advances in physics, mathematics, and other sciences have given rise to theories spanning the world of particles to the depths of the cosmos. Theories like Einstein's relativity, quantum mechanics, and the Standard Model have greatly contributed to our understanding of the universe's building blocks. However, none of these theories have succeeded in providing a unified framework—a "Theory of Everything"—capable of explaining all aspects of the universe.

This book chronicles a journey toward a theory that offers a comprehensive understanding of the universe's workings, explaining all physical phenomena within a single framework. The Theory of Everything encompasses a vast spectrum, from the universe's most fundamental building blocks to large-scale structures and processes. This theory is not merely a product of scientific curiosity; it also symbolizes the human mind's response to profound questions.

I invite you to join me on this journey of discovery. Together, we will unravel the universe's mysteries and uncover the fundamental principle behind everything. Venturing beyond known boundaries, we'll embark on a scientific adventure to explore the enigmatic fabric of the universe. As we delve into the Theory of Everything, we'll gain fresh perspectives on both the universe and ourselves.

Nothingness is one of the deepest and perhaps most complex concepts in human thought. As the antithesis of existence, representing emptiness and absence, nothingness has captivated philosophers, scientists, and artists throughout history. Reflecting on nothingness means confronting the most fundamental questions about existence and the universe.

In the Theory of Everything, I use nothingness not as the absence of matter but as the singularity of matter. For example, when someone in a desert says, "I'm alone in the middle of nothingness," they refer to a nothingness made of sand; someone at sea saying, "I'm stuck in the middle of nothingness," refers to a nothingness composed of water. In these examples, I emphasize that the desert and the sea are singular environments defined as nothingness.

Before the Big Bang, space, and the universe, absolute nothingness was an infinity composed of the essence of matter—not in particle form, but as a single, singular entity. The moment this eternal absolute nothingness loses its absoluteness, an everlasting nature begins to form.

DEFINITION OF PHYSICAL ENTITIES

In the Theory of Everything, I define existence in nature based on three conditions, all of which must be present simultaneously in anything that exists. (Remember, absolute nothingness is not the nature we live in; it exists before and beyond nature!)

Conditions for Existence in Nature

1. To Exist, Something Must Form

Explanation: Everything in nature is composed of micro-particles. The interactions, shapes, and volumetric density changes of these micro-particles give rise to a multitude of objects and events at the macro level. Change, formation, or transformation occurs as a result of the reactions substances exhibit to interactions from other substances.

2. To Exist, Something Must Occupy a Volume

Explanation: Occupying a volume defines the boundaries of the space a physical entity or cluster of entities takes up. In other words, everything we define as existing in nature has a finite volume and boundary. Nothing that exists in nature can be infinite in size, number, or extent.

3. To Exist, Something Must Move

Explanation: Due to the second condition, entities with the same or different volumes and densities interact and react upon contact. These interactions among entities generate pressure and fluctuations. This pressure force is fundamentally positive, and as a fundamental force, it gives rise to negative forces. Consequently, every cluster of entities must move, both internally and as a whole.

Everything in nature possesses these three characteristics, which are interconnected.

Addendum: What Are the Conditions for Interaction and Reaction?

For an interaction to occur and a reaction to take place in nature, at least two entities must come into contact. For a reaction to occur between contacting physical entities, two conditions are necessary:

First Condition: For entities in contact to interact and react, they must have different properties (including symmetry or asymmetry, volumetric size, and density).

Second Condition: If two symmetrical physical entities with identical properties come into contact without a third force, no reaction will occur. Therefore, there must be at least four entities with identical properties in contact. In a cluster of entities with the same properties, the first reaction will only be inward or outward pressure. Since everything is composed of matter, every substance is at least in a state of vibration.

Note: I explain these concepts in the introduction to clarify what I mean by "nothingness." When I mention nothingness, many people think of non-existence. However, I believe that understanding what existence entails will make it easier to accept the impossibility of non-existence.

To make this clearer, let's conduct a thought experiment within the space system, here on Earth. Imagine two vacuum tubes of the same size and properties, both filled with air. Since this is a thought experiment, let's assume the tubes are indestructible. Now, let's vacuum one of the tubes. What is inside the vacuumed tube that remains intact? Since the tube has been vacuumed, its occupied volume remains the same. By removing the air—i.e., the known elements—either something was already inside the tube, or something new formed due to the vacuuming. The vacuumed tube has the same volume as the non-vacuumed one, but its density has significantly decreased. As you can see, the volumetric space inside the vacuumed tube meets all the conditions for existence: it has formed, occupies a volume, and is in a state of vibration.

In this thought experiment, we transformed the airless space inside the vacuumed tube into a region with reduced air density. Thus, by completely vacuuming out the elements, we find dark matter inside the tube. If dark matter weren't already present, wouldn't something new have been created from nothing? Wouldn't energy or matter created from nothing increase the efficiency of that system?

In our experiment, vacuuming the tube removed the asymmetrical elements between symmetrical dark matter clusters. In an isolated system like space, it is impossible for a void (absolute nothingness) to exist or form between points A and B. Therefore, non-existence is impossible in isolated systems. Similarly, non-existence is impossible in non-isolated systems, which are boundless. In a boundless system, the concept of infinity emerges. Since infinity has no end, such a system has no beginning either. In other words, infinity—or absolute nothingness (singularity)—has existed eternally.

ABSOLUTE NOTHINGNESS

I previously defined existence and explained the impossibility of absolute nothingness between two points in space. Therefore, absolute nothingness does not possess the characteristics of existence as I defined them. Being infinite, absolute nothingness is unformed and has no beginning; it lacks a volumetric space and is temporarily static. As you can see, absolute nothingness has none of the traits we associate with existence in nature. However, we must remember that static states are temporary and lack continuity.

1. Absolute Nothingness Is Eternal: It is unformed and has no beginning.

2. Absolute Nothingness Is Infinite: Infinity cannot be defined volumetrically because it has no boundaries.

3. Absolute Nothingness Is Temporarily Motionless: Due to its infinity, it is temporarily static.

Note: To imagine the concept of infinity, you must never think of an end. If you imagine an end, you're thinking of a finite cluster, not infinity. Accept the concepts of "eternal" and "infinite" as they are! Imagining a beginning in eternity or an end in infinity leads to illogical thinking, which can exhaust your mind.

Absolute Nothingness Is Singularity

Singularity: The state of being singular, a single entity where every point shares the same properties.

As you can see, I've defined singularity in a very simple way.

The reason we cannot define absolute nothingness—or singularity—as existing is the absence of something different to compare it to, as nature has not yet formed. Due to the impossibility of non-existence, absolute nothingness physically exists. However, being infinite, it is dominated by uncertainty. The structure of absolute nothingness, or singularity, is identical to that of all matter in nature; the difference is that it consists of a single piece rather than countless micro-particles. Absolute nothingness is the essence of matter. Since it is unformed and eternal, it is not in particle form. We can also describe it this way: Because non-existence is impossible even in infinity, absolute nothingness has existed as a singularity since eternity. It is temporarily static, with the properties of point A being identical to those of an infinite number of points B. Since there are no entities within absolute nothingness, most physical laws defined by humans have not yet emerged. The only state in absolute nothingness is the uncertainty arising from its static condition. The only thing that can disrupt a static state is change, and every change involves the simultaneous action of positive and negative forces.

Note: Positive force is a pushing force; negative force is a pulling (vacuum) force.

For example, if you push a car to move it, you apply a positive force; if you pull it, you apply a negative force. When pushing the car, your mass and muscles generate an inward-to-outward positive force. When pulling, the same inward-to-outward positive force is produced. Positive force is primary, while negative force is secondary, and they always emerge simultaneously in everything, in every isolated system. In an isolated system like space, where every area is filled with atoms, positive and negative forces always occur together. Thus, the effect of positive forces generates reactive negative forces, leading to a chain reaction where an action produces a reaction, and the reaction produces a new action.

Positive force creates tension, which can release singularity or absolute nothingness from its static state. This tension, caused by positive force, can lead to a tear at any point in absolute nothingness. To illustrate tension: If you take a thin rope you can break and pull it from both ends with your hands, you apply positive force in opposite directions. This causes tension in the rope, as the rope tries to return to its original state while you continue to stretch it. The rope's effort to return to its original shape and your force to break it momentarily reach a balance. At this moment, the rope will either break or not; this is when uncertainty arises in the atoms composing the rope. This moment also marks the limit of the rope's tension. If the positive force prevails, the rope breaks. In absolute nothingness, if the positive force causing tension prevails, a tear occurs at some point due to infinity's countless directions. You can also view this tear as a result of the change induced by the positive force acting on the static singularity.

Important Note: For those struggling to imagine tension in absolute nothingness, you can accept that the first interaction in absolute nothingness is change. Change is an undeniable reality in our lives and in nature. According to causality, change is the first interaction in absolute nothingness.

BEFORE THE BIG BANG

CHAPTER 1

What was the first change in the temporary static phase of absolute nothingness or singularity? We cannot know or falsify this, but according to causality, one of two possible changes may have occurred: either solidification or softening.

As we know, phenomena like heating, burning, freezing, or cooling are perceived macro-level natural events. When we set perception aside, at the micro/quantum scale, matter in nature exists in only two fundamental states: solidification and softening. You can observe your surroundings if you'd like; everything around you exists at some level of solid or soft states. In nature, matter (atoms) exists in solid or soft forms, independent of perception. The macro-level states of matter we witness in nature are merely the perceptual reflections of events caused by micro-particles (elements and photons) detaching from dark matter.

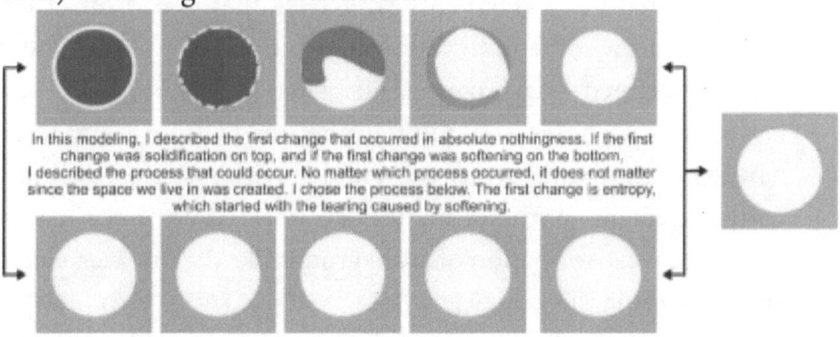

If the first change in eternal absolute nothingness was solidification, the event described in the upper part of the model above occurs. Since singularity, or absolute nothingness, is infinite, solidification begins simultaneously at an infinite number of points. During solidification, a solidifying point exerts a negative force on another solidifying point, resulting in the simultaneous formation of an infinite number of solid atoms. At the same time, soft atoms form through tearing. If this hypothesis holds true, an infinite number of spaces would form simultaneously within infinity, potentially leading to the creation of countless universes, similar to the existing multiverse theory.

In infinite singularity, the hypothesis in the lower part of the model involves a softening change. When infinite singularity begins to soften, the resulting tension leads to a tear at a single point in infinity, forming the exterior of the space we inhabit. I will call the first cluster of objects that form "photons," as I previously defined soft atom clusters as photons. However, this photon is larger than space, and space will later form within this cluster.

As a result of the change in absolute nothingness or infinite singularity, the first photon—a cluster of soft, singular atoms—emerges. The fundamental structure of this massive photon is the same as that of singularity. However, during the tear, just as crumbs fall when you break a stale piece of bread in half, asymmetrical fragments detach from singularity, the essence of matter. Since these fragments stretch as they break off, a cluster forms from the soft and flexible state of atoms. Thus, from the three classes I defined—dark matter, elements, and photons—a photon has formed. A note: These initial photons will not be part of our nature, meaning they won't be included within space. Additionally, neither elements nor dark matter have formed yet. Keep reading; I will explain the formation of atom groups like dark matter, elements, and photons in due course.

As shown in the diagram below, there is no issue with naming nothingness and the pure energy field (photon) with different terms.

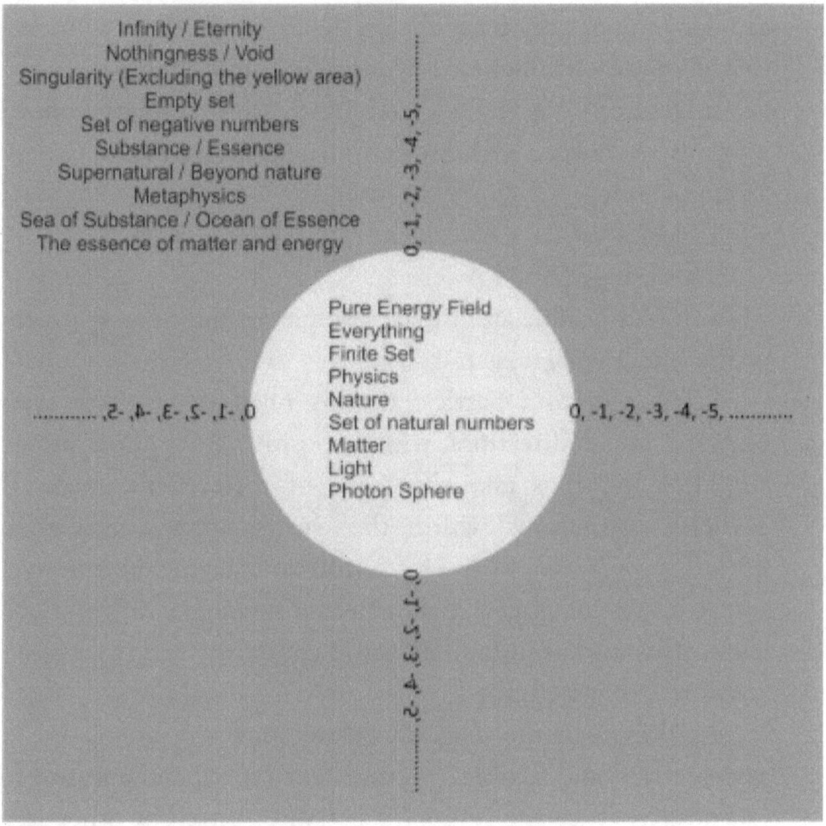

After this initial event, absolute nothingness loses its absoluteness. With the onset of entropy, absolute nothingness becomes impossible. Consequently, our previous definition of "eternal absolute nothingness" now transforms into "absolute nothingness is impossible and eternal." The entropy that begins in absolute nothingness or singularity allows everything to exist in place of nothing. This also answers the question, "Why is there something rather than nothing?"

Chapter 1 Summary

Due to causality, the first change in singularity could be either solidification or softening. Regardless of which occurred, it cannot be falsified; however, since our reference point is "change," no one can deny the pre-Big Bang change dictated by causality. We learned that absolute nothingness (an infinitely large, single-piece matter, i.e., singularity) was eternal, but with the formation of a photon or photons, the onset of entropy transformed absolute nothingness into an eternal state.

CHAPTER 2

Clustering of Solid Atoms

Soft and flexible atoms have completely filled the space created by the tear, leaving no gaps.

Each soft atom particle in this filled space exerts pressure outward in all directions with a positive force. Soft atoms are shapeless and thus take the shape of their environment. Being particles, soft atoms within the cluster vibrate, generating an inward-to-outward positive force. You can imagine the movement of soft atoms in this space like the constant shaking of different-colored jellies. This is what sustains entropy's continuity and makes reaching absolute zero impossible.

As shown in the diagram below, each soft atom, due to its asymmetrical and flexible structure, tends to expand outward with a (+) positive force and resists the pressure exerted by other atoms it contacts.

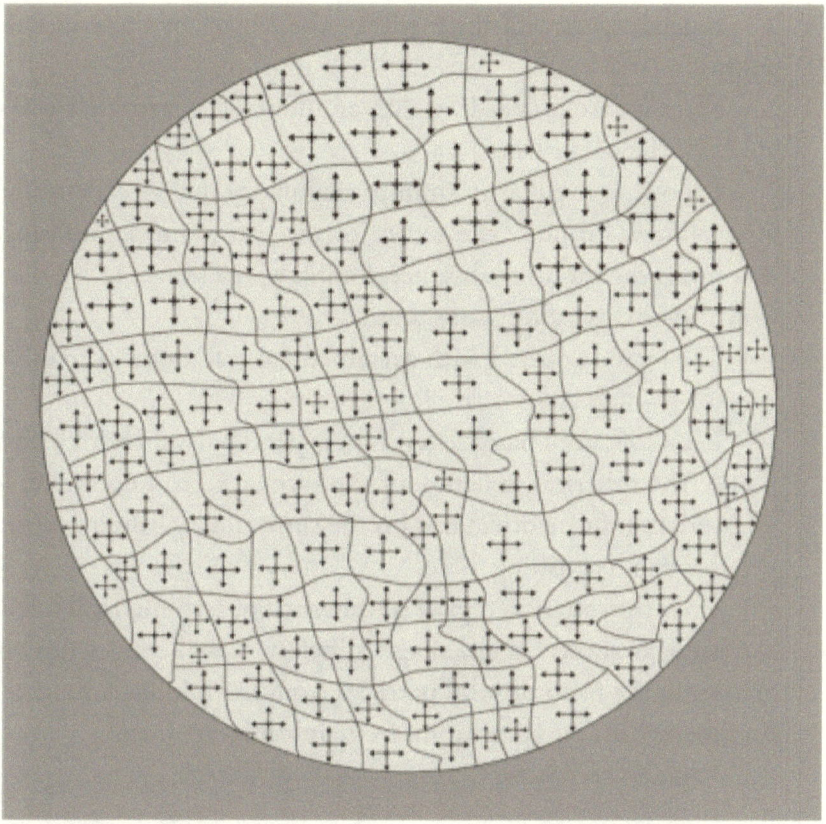

The region formed by soft atoms is dynamic due to the third law of thermodynamics. The reason matter cannot remain still without vibrating lies in the fact that each atom—the smallest indivisible soft particle—has its own volumetric boundary. When these asymmetrical entities with defined boundaries come into contact, their outward positive forces differ, creating a force imbalance within the cluster. This imbalance keeps the soft atoms in constant motion, much like a jiggling jelly.

There is no difference between the cluster of soft atoms and the light particle clusters in our nature; both are clusters of soft and flexible atom particles. However, the region of the first soft atoms to form is much larger. Here's a hint: space exists within this region.

Indeed, space will form within the cluster of these initial soft atoms.

Since we are proceeding with an inductive approach, I will now describe the next event.

Following the tear in infinite singularity due to expansion, the first photon formed. The source of the soft atoms composing the photon is the infinite substance. Similarly, the source of solid atoms will also be this infinite substance.

The region covered with asymmetrical soft atoms contains only inward-to-outward positive forces.

Each soft atom particle tends to expand outward with a positive force. The spherical structure of the formed region exerts immense pressure in infinite directions on the surrounding infinite substance sea through its pushing force.

Since the spherical region is composed of asymmetrical soft atom particles, as shown in the diagram below, it will extract indivisible masses in the form of solid particles from the infinite substance sea. In other words, singular atoms in their solid state form...

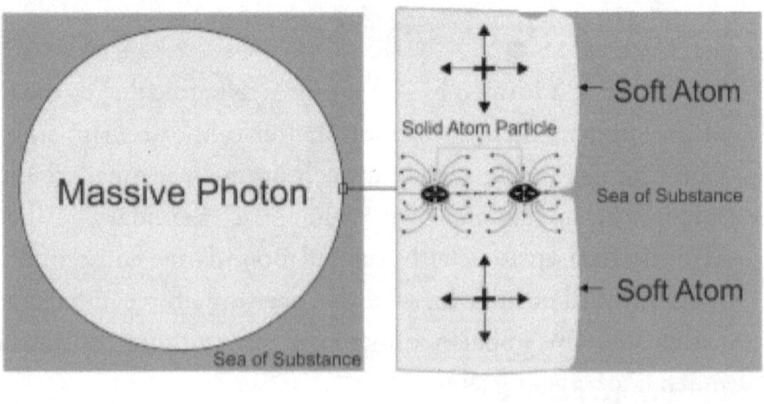

As shown in the magnified view in the diagram above, the soft atom particles, depicted in yellow, exert pressure on the sea of substance, causing small solid particles to detach from it. These particles, shown in black in the diagram, are solid atoms.

A solid atom particle is metaphorically equivalent in weight to a soft atom particle. The reason a solid atom is smaller than a soft atom is that it has not expanded—or, in other words, it has not been subjected to sufficient negative pressure to stretch like rubber and increase in volume.

This event can be analogized as follows: Imagine a container filled with dough. The dough represents the infinite substance, and your fingers represent the soft atom particles. When you apply positive force by pressing on the dough, even if your fingers are close together, the dough will pass between them. Picture a similar mechanism, but applied in three dimensions, from inside a spherical structure to the infinite sea of substance outside. In another analogy, imagine yourself trapped inside a balloon, pushing it outward with your hands. The inner surface of the balloon emerges between your fingers, but solid atoms pass through in tiny fragments. Thus, following the formation of soft atoms, the first solid atoms begin to form.

So, how do these solid atom particles reach the center of the spherical pure energy field?

The mechanics of gravity and mass attraction have long been among the most thought-provoking topics for humanity. While many accept gravity, they remain skeptical of the theories explaining its mechanics. As I mentioned earlier, since we are proceeding inductively, I must also explain how physical phenomena arise. In the Theory of Everything, there is no fundamental force in nature called gravity or mass attraction. I will elaborate on this proposition in later sections. However, I will use the terms gravity and mass attraction metaphorically, as they describe a natural phenomenon. The mechanics of gravity emerged before the Big Bang. Since there is no complete void anywhere, the idea of matter attracting matter is neither possible nor logically sound. All phenomena we define as gravity and mass attraction at the macro level fundamentally occur through positive force. You can test this as many times as you like—the result will not change. At the micro level, every positive force you create, naturally or artificially, will inevitably produce a negative force. For example, when you hold an object and pull it toward yourself, it may seem to demonstrate negative force; however, the first thing that occurs is the positive pressure effect of blood filling the capillaries in your muscles from your veins, similar to a hydraulic system in vehicles.

Gravity: At the quantum level, it arises from the resistance and positive pressure relationship between asymmetrical particles. Whichever direction the positive pressure is stronger in, matter moves in the opposite direction or toward the area of weaker pressure. This phenomenon has been defined as gravity at the macro level. You will understand this event better later on.

Mass Attraction: While its behavior resembles gravity, the primary factor is the magnetic field effect of the object at the macro level. In magnetic fields, the primary force is also positive force vectors. At the micro level, mass attraction is related to geometric shapes. I will explain gravity and mass attraction in more detail later.

As shown in the diagram below, the force distribution in the photon's field is in all directions, primarily inward to outward, due to the presence of soft atoms. The reason solid atoms move toward the center is that the positive force detaching the solid atom from the sea of substance is stronger in the soft atom particles at and near the outer surface. The space occupied by each solid atom entering from the outside increases the pressure level on the outer surface of the photon. Additionally, the large volume of soft atoms and their vibrational motion are separate contributing factors.

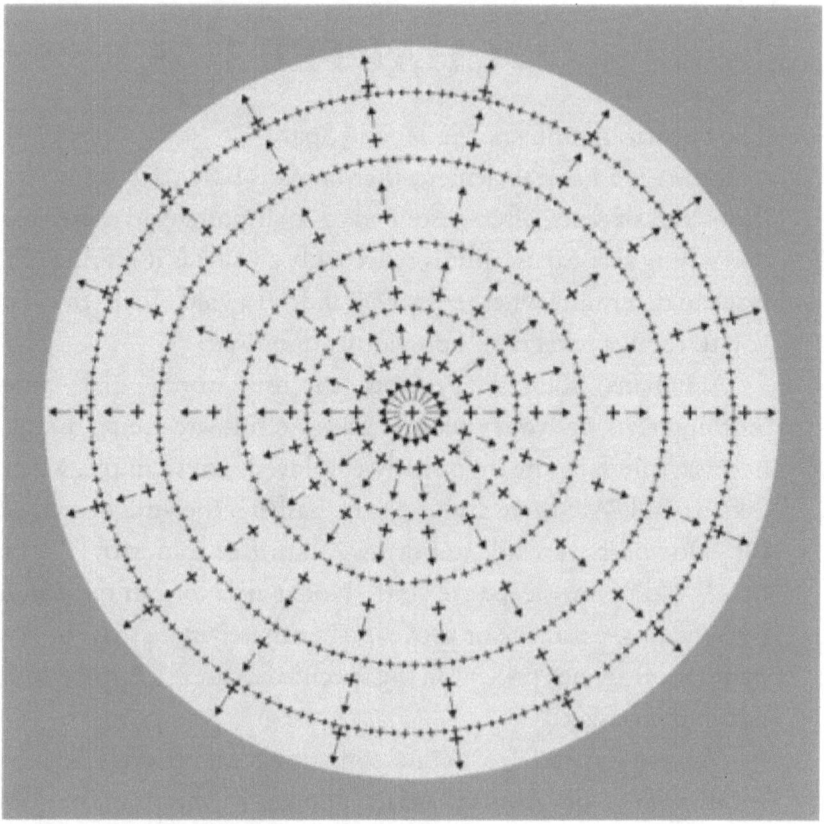

CHAPTER 2 SUMMARY

When infinite singularity (the sea of substance) tears due to expansion, the photon cluster formed by the initial soft atom particles, being asymmetrical, exerts uneven positive force on the elements it contacts. The pressure generated in all directions pushes the solid atoms detached from the sea of substance toward the center of the sphere. Thus, the phenomenon we now know as "gravity" occurs for the first time.

Gravity is the tendency of matter, compressed by positive force from all directions, to move toward the direction with the least force resistance. (I will revisit gravity later.)

CHAPTER 3

The Formation of Dark Matter and Space

Space is a natural element that humans have tried to unravel for thousands of years, often described as "an infinite void encompassing everything that exists." But is space truly a void? Is it infinite? First, I want to determine whether space is indeed a void. To do this, I'd like you to recall a paragraph I previously discussed:

"Humans sometimes need to use words like 'absence,' 'nothingness,' or 'void' because these terms are quite useful and necessary in daily life, frequently employed. For example, when you ask a friend, 'Are there apples in the basket?' they might reply, 'No.' Or, 'Do you have work tomorrow?' 'I'm free tomorrow.' 'Is Hasan here?' 'No.' 'How sleepy are you?' 'Not at all.' None of these indicate physical non-existence or prove it; they merely state that the entities in question are not present in the specified context. (Apples exist, but not in this basket.)"

As you can see, to define something using terms like "void," "nothingness," or "absence," we need to know what exists or does not exist. For instance, to the question "Is there a universe in space?" we answer, "Yes, there is." To "Are there galaxies in space?" we also say, "Yes, there are." Now let's change the question and assume the universe consists of only two galaxies: "What is between the two galaxies?" The answer becomes, "A void." Here, the response changes; we have created a void. In our universe of two galaxies, there are no other galaxies, so we can define the space between them as empty due to the known element of "galaxies," thus physically establishing the void. This shows that even if we think of space as an empty set without the universe or light, the non-existence of space is impossible. When we consider space as an empty set, it becomes clear that space is finite. Recall the three conditions for existence: 1) Formation, 2) Occupying a volume, 3) Movement. If we define space as either an empty set or as existing, it must fulfill all conditions of existence. Therefore, space is a finite set that has formed, occupies a volume, and is in a state of movement or change.

With this understanding, I can now explain how space formed, how it gained volume, and how it remains in a state of change.

The inward-to-outward positive pressure force created by the soft atoms covering the massive photon sphere causes the solid atoms detached from the infinite sea of substance to cluster at the center of the spherical pure energy field. These solid atoms, breaking away from the sea of substance, gather at the center of the pure energy field like rain falling to the ground. During this process, mass is continuously taken from the sea of substance. Each wave or swarm of solid atoms is exposed to greater duration and stronger pressure than the subsequent wave, leading to the formation of sub-clusters within the main cluster due to the distance between waves. Each new sub-cluster that emerges is a newly formed dark matter. The number of dark matter clusters, which are groups of solid atoms, will multiply exponentially with each wave.

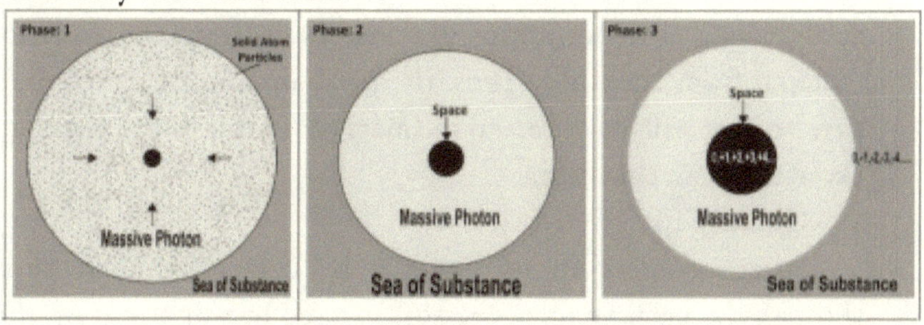

In the diagram above, the imaginary line included in the infinite sea of substance in Phases 1 and 2 represents the potential total volume of the space set in the next stage.

Space is always a finite set. Since solid atoms continuously detach from the sea of substance, our space set will grow indefinitely. Therefore, the mass within the area enclosed by the imaginary line between the massive photon field and the sea of substance represents the potential mass the space set will gain.

In Phase 3, I used the number 0 at the boundary because zero represents the boundary of a set. The boundaries of a set become part of the set as it physically grows or shrinks. Thus, zero has a physical counterpart. For example, let's consider an apple as a set, with its peel as zero. When we eat the apple with its peel, we also consume the boundaries of the apple set.

To illustrate this with an analogy: Imagine entering a cave with no exit to dig, and the entrance collapses behind you. You are now inside the cave. The area within the cave represents the massive photon field, and the walls trapping you, including the collapsed entrance, represent the sea of substance. You are trying to dig your way out with a pickaxe. In this scenario, you resemble the soft atom particles of the massive photon field, extracting matter from the cave walls. Each piece of material you extract with the pickaxe represents solid atom particles. You then gather all the extracted material at the center of the cave. As you can see, every piece of material accumulated at the center of the cave results from the depletion of material from the walls. Thus, the total amount of material at the cave's center will always equal the mass or volume of the material you extracted from the walls.

CHAPTER 3 SUMMARY

Space acquires all its material by detaching small particles through the positive force pressure exerted by soft atoms on the sea of substance. The massive photon, formed at the moment of the initial tear of singularity or absolute nothingness, is a field where this transition function occurs.

Since these actions will repeat continuously and indefinitely, space and the massive photon field will grow proportionally, gaining mass and volume. The solid atoms accumulating at the center of the massive photon become components of newly formed dark matter.

Due to the pressure force differences over time, dark matter forms symmetrical particle clusters. The new cluster formed by dark matter is space. The total volume and mass of the space set are equal to the mass and volume lost by the sea of substance.

CHAPTER 4

How Did the Big Bang Occur at the Center of the Space Set?

The space set, a cluster of dark matter formed within the spherical massive photon field, will reach enormous sizes as it grows. As space expands, the pressure exerted by the massive photon field outside it on the space set increases, causing the dark matter composing the space set to be continuously compressed. It is evident that dark matter, compressed with constant and increasing force, can never remain still (without vibration). Moreover, dark matter resists the external-to-internal pressure by exerting internal-to-external resistance.

As shown in the image below, the massive photon sphere is positioned between the space set, composed of dark matter, and the infinite sea of substance. Consequently, the resistance and positive pressure force exerted by the pure energy field on both space and the sea of substance will continuously increase in a proportional manner. This also means that the massive photon sphere will extract matter (solid atoms) from the sea of substance more rapidly and in greater quantities.

If you have grasped that the dark matter composing space consists of the smallest volume atoms (compressed or unexpanded in their raw state), that the massive photon sphere is a transition field made of indivisible soft atoms (which expanded and became soft during the tear), and that the infinite sea of substance is the foundation of everything, I will no longer need to mention the massive photon sphere and the sea of substance.

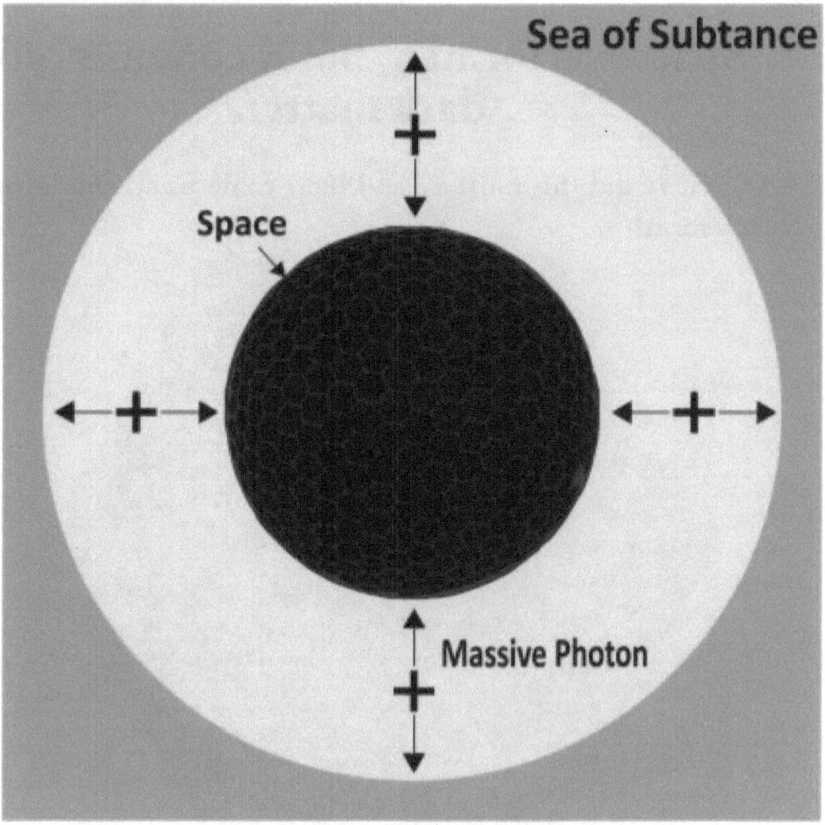
Sea of Subtance
Space
Massive Photon

Space has formed, but it contains only dark matter.

So, where did this universe and light come from, and how did they form?

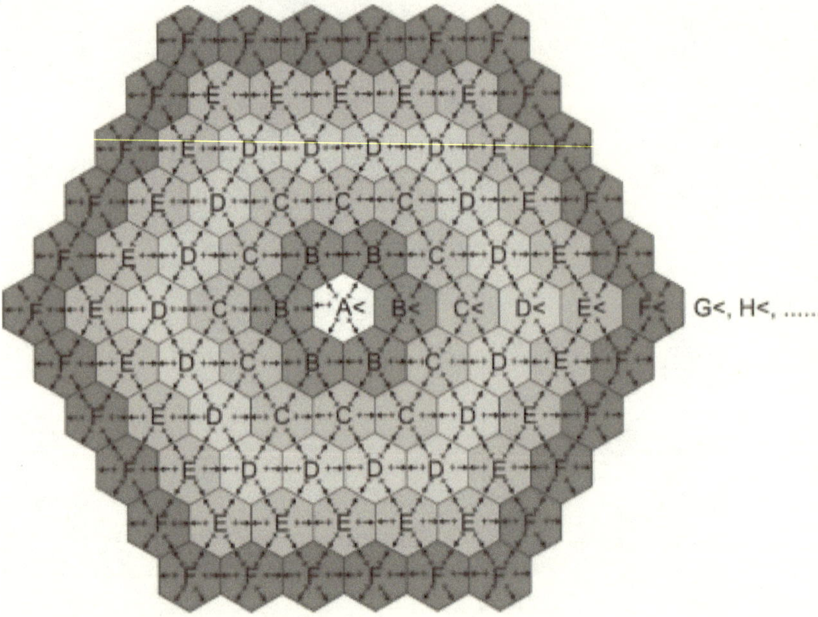

In the model above, you can see the hexagonal-cubic dark matter structures at the center of space.

In the image, the A at the center represents the most primordial (first-formed) dark matter in the center of space.

For this reason, it has been resisting the positive force pressure from the outside inward for a long time. From another perspective, since it is the closest to absolute zero in space, it is the dark matter in the most solid state.

On the Concept of Symmetry

In this book, I use symmetry to mean "sameness." For example, imagine two apples with approximately the same size. These apples are symmetric. Now imagine two apples again, but this time their size, color, weight, and even the number of atoms composing them are exactly the same. I define these apples as "super symmetric." Additionally, consider two baskets, both containing apples of various sizes, but the baskets have the same weight. These two baskets (sets) are symmetric.

Therefore, in terms of volumetric space, the smallest dark matter is A.

If the A dark matter particle is the smallest, then the dark matter particles in the B sequence are slightly larger than A.

Since the A dark matter and each dark matter particle in the B sequence are only slightly different from each other, A and B are symmetric. However, the dark matter particles within the B sequence are super symmetric relative to each other.

As a result, we can define it as follows: From smallest to largest, A < B < C < D < E < F < G < H < ... This sequence continues outward to the edge of space.

For everything to be super symmetric is quite difficult due to the dynamics of the action-reaction formation we discussed earlier. Since absolute nothingness is impossible, every unit in nature must come into contact with another unit. Consequently, each unit exerts pressure force on the other. Each unit subjected to pressure resists this force as a requirement of existence. Because the force exerted by multiple units on others varies, it becomes impossible for all natural units to remain in a super symmetric state. For example, in the spherical space set, dark matter near the surface and dark matter near the center are not subjected to the same level of pressure. Therefore, particles near the surface occupy significantly more volumetric space compared to those near the center. In other words, while those near the surface are softer, those at the center are more solid. Similarly, since the particles near the surface have a greater range of motion (vibration), their metaphorical temperature is higher; those at the center have less range of motion, so their temperature is lower. Thus, the dark matter at the center has come very close to reaching absolute zero. (I will discuss heat and temperature later.) I refer to dark matter as "particles," but each of these dark matter entities is a cluster of solid atoms.

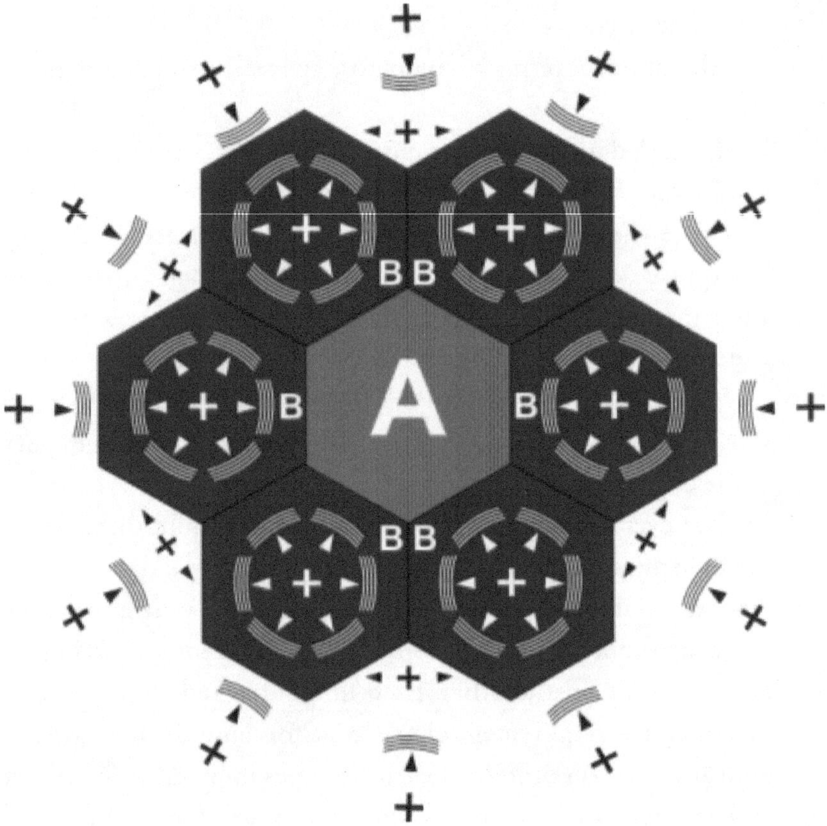

In the diagram above, the dark matter (A) at the center of space bears the entire load and pressure of space. Its range of motion has become extremely limited, and its resistance point is on the verge of breaking. However, the resistance point has not yet broken because the resistant matter particles in the B sequence exert force in all directions. You can compare the way the B sequence and other sequences reduce the force experienced by the inner sequence to the structure of a mosque dome, an Eskimo igloo, or an arch-shaped bridge.

You may be familiar with these mechanisms through physical concepts such as Load Distribution, Stress Distribution, Shielding Effect, Pressure Transmission, Arch Effect, or Dome Effect. If we also consider the C, D, E, and other sequences, it is clear that the symmetry in space cannot be disrupted until space reaches enormous dimensions. However, when the size of space and the pressure it experiences from the outside inward increase proportionally to a massive scale, a gap will form between the A dark matter and the B dark matter sequence.

The dark matter particles forming the B sequence create a negative force pressure (vacuum due to the arch effect) toward the A dark matter, resulting from the positive force pressure experienced from the outside inward. Due to this negative force pressure, the solid atoms on the surface of the dark matter begin to stretch. As the solid atoms stretch, they start transforming into soft atoms. You can think of this negative force pressure or vacuum as the attraction between two sticky objects when you try to separate them, because absolute nothingness or void is impossible. Therefore, the dark matter particles in the B sequence, which are in contact with each other, and the atoms on the surface of the primordial A dark matter at the center begin to soften and stretch due to the vacuum effect. As the solid atoms stretch, they transition into a softer and more slippery state. It is extremely difficult for atoms that transition from a solid to a soft state through stretching to return to their original form due to entropy. You can imagine this as trying to push cream back through a perforated strainer or pushing toothpaste back into the tube after it has been squeezed out.

As the solid atoms on the surface of the central dark matter stretch and transform from a solid to a soft state, the arch effect of the B sequence and the super symmetric dark matter sequences above it breaks. Consequently, the first spark forms at the center of space. The super symmetry of the central dark matter is now disrupted.

This spark (a metaphor here; by spark, I mean the transformation of a large number of solid atoms into soft atoms), or photons, slides outward chaotically through the dark matter particles in the B sequence, breaking the symmetry. This event triggers a chain reaction, meaning the Big Bang occurs.

From this point onward, you can imagine the cluster of soft atoms, or photons, as air bubbles under the sea, and space, composed of dark matter, as the sea itself.

The photons formed during the Big Bang will move toward the surface of space, much like air bubbles rising to the surface of the sea—or more precisely, from a high-pressure area to a low-pressure area.

If you examine the diagram below, you will see a cluster of photons made of soft atoms trapped between symmetric dark matter sequences. Each dark matter particle that detaches from its sequence and enters the soft atoms is subjected to the pressure of the soft atoms, causing it to slide and accelerate in random directions. During this acceleration, it may collide with another dark matter particle or even with other dark matter sequences whose symmetry has not yet been disrupted. Due to the force of these collisions, the dark matter undergoes further fragmentation (chain reaction). After the collision, some dark matter particles break apart, resulting in the formation of asymmetrical, smaller particles. You can imagine this as two symmetric stones colliding and producing asymmetrical smaller stones.

During the countless collisions in the Big Bang, numerous dark matter particles fragmented, losing their symmetry, and countless asymmetrical elements emerged.

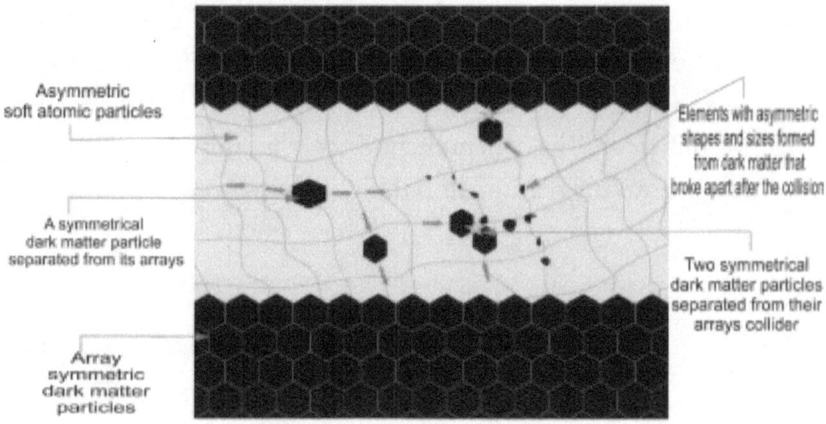

After the series of events that triggered the Big Bang, the symmetric dark matter sequences at the center of space reform. Since the entire load of space continues to accumulate at the center, dark matter particles that have not lost their symmetry take their place at the center.

The symmetric dark matter particles that regroup at the center push the elements and photons formed due to the pressure they exert on each other in all directions outward from the center. (Since we consider photons as air bubbles and space as the sea, we can think of elements as objects within the sea.)

Just as creatures or objects underwater rise to the surface with the help of air bubbles, photons carry elements to the surface of space or to regions with less pressure.

CHAPTER 4 SUMMARY

The strong positive force from the outside inward acting on the center of the spherical space causes the symmetry of the central dark matter to break. When symmetry breaks, atoms that stretch and detach from the dark matter create the first spark, leading to the Big Bang. With the Big Bang, a new entropy begins, and through countless chain reactions, numerous elements and photons form in space. Dark matter particles that retain their symmetry start clustering again at the center of space. This clustering pushes all elements and photons outward from the center. Thus, space is no longer a void but a home for the countless newly formed elements and photons. The phenomenon of the universe's expansion is analogous to the sea and air bubble analogy.

CHAPTER 5

Introduction to Dark Matter, Elements, and Photons

Let us redefine the contents of space by dividing them into three main classes. These three main classes can further be divided into subclasses.

Everything in space, from its initial formation to the present, is composed of indivisible atoms. Humans, other living beings, mountains, seas, lakes, oceans, planets, stars, galaxies—everything we know or don't know—is fundamentally made of atoms.

Atoms are the smallest, indivisible fundamental building blocks. Atoms have two extreme states: the first is their raw state, which is solid but small in volume and mass; the second is their soft state, which is larger in volume. Between these two extremes, there are countless intermediate stages of transition from solid to soft.

1. Dark Matter

Dark matter is a cluster of compressed atoms in a solid state. Nearly the entirety of space (a ratio that mathematicians can or may have calculated) consists of symmetric dark matter. Dark matter and dark energy are the same. However, we will refer to dark matter in areas with a high presence of asymmetric photons and elements as "dark matter," and in areas with little or no presence of them as "dark energy." We cannot directly see dark matter, but we can feel and observe the phenomena resulting from its effects. For example, bring your index finger and thumb about 1 millimeter apart and look at the 1-millimeter gap while pointing at any spot. The light refracts, and an aura or magnetic field lattice wall becomes visible on your skin. This is the effect of dark matter. The attraction or repulsion of magnets is also a dark matter effect.

However, it is impossible to directly see dark matter itself because photons move between dark matter particles. From another perspective, we can say that everything dark is dark matter. If we see light and everything else through the influence of photons, we can say we see darkness through the influence of dark matter. Dark matter is entirely composed of atoms in a solid state. However, if dark matter is subjected to excessive friction due to passing photon clusters and vibrating elements, the atoms on its surface soften at the tips while their roots remain solid. (This will become clearer in the hypotheses you will read later.) Additionally, if dark matter is caught between a large photon cluster, it may detach from its sequence, accelerate through the photons, and collide with another dark matter particle, causing fragmentation. Some situations where dark matter fragments include phenomena like nuclear bombs, supernovas, the Big Bang, solar flares, lightning, thunderstorms, and volcanic eruptions. When dark matter fragments, asymmetrical elements emerge.

2. Elements

Elements are the fragmented or broken state of dark matter. An analogy for this is the breaking of a glass cup: when a glass cup hits something at sufficient speed, it shatters, and a single cup produces dozens or even hundreds of asymmetrical glass shards. Similarly, when dark matter fragments or breaks, asymmetrical elements emerge. Humans have become highly skilled at identifying elements and have created the periodic table. Since this does not significantly contradict the Theory of Everything, I have named all asymmetrical solid atom clusters and particles that result from the fragmentation of dark matter as "elements." In the Theory of Everything, when dark matter breaks, new asymmetrical particles are formed, and humans have already classified these asymmetrical particles. Therefore, I found it appropriate to use the term "element." However, in the Theory of Everything, the internal structure of elements is different; unlike traditional atomic models, there are no quarks, positrons, neutrons, or protons. But don't worry! The Theory of Everything can easily explain all the experiments and natural phenomena that other theories can account for. What truly matters is the systematic definition of events and entities in nature.

Since dark matter is composed of highly compressed atoms, the elements that emerge when it breaks are also compressed and in a solid state. Some of these elements, after fragmentation, naturally come together again with others of their kind due to micro and macro disturbances. For example, asymmetrical cube-shaped elements may come closer through disturbances, reattach, and form something like gold. Humans have made significant advancements, from medicine to food, by first separating and then combining elements. An analogy for this is as follows: Imagine you have a garden with onions, peppers, and tomatoes. When you pick the onions, peppers, and tomatoes from the garden, you separate them; when you chop them into small pieces and put them on a plate, you combine them, and then you enjoy your salad. Scientists also separate elements while collecting them from nature or afterward and then combine them through various processes. This is how our technology advances, and we benefit from its fruits. I hope you will better understand elements and their model in the Theory of Everything through the hypotheses in later sections.

3. Photons

Photons are the source of countless benefits, including light, heat, and communication, while also being mysterious. The Theory of Everything unravels all the mysteries of photons.

Photons are clusters of soft atoms. They form when a solid, compressed, and small-volume atom in its raw state expands under negative pressure (like stretching a rubber band in opposite directions), becoming larger in volume, loose, and soft, and then clusters together. In the Theory of Everything, soft atoms are defined as electrons when close to an element and as photons when not close to an element.

Each photon is a fluid, asymmetrical, and shapeless cluster made up of an unknown number of soft atoms. The cores of stars and planets are formed by photon clusters. Photons can travel at the speed of light but can also remain relatively stationary in a given space at times. I can almost hear you asking, "How can this be?" Photons are made of clustered soft atoms, and these clusters exhibit behavior relative to their size (the number of soft atoms they contain) and the pressure they are subjected to. For instance, a cluster made of a few soft atoms (an electron) can remain in our bodies for years—perhaps even electrons that have been with us since birth. Most photons, however, continuously travel at the speed of light until they collide with something and disperse. Some photons remain intact for billions of years in the cores of planets and stars. For example, neutron stars are actually giant photons. Photons forming the cores of planets and stars, as well as those labeled as neutron stars, maintain their structure unless a supernova occurs. Photons traveling at the speed of light are neither very small nor massive clusters. For photons moving at the speed of light, mathematicians and physicists will have a lot of work to do; I can only provide examples.

The size of photons determines the frequency of light. Since photons are situated between dark matter particles, they create a wavelength relative to their size. I can explain where photons get their speed and how it is determined with the following analogy: Everyone has handled a wet bar of soap with wet hands. Imagine your hands as dark matter and the soap as a photon! When you apply positive force pressure to the soap with your hands, it slips away, doesn't it? In nature, the movement and acceleration source of a photon is similar. The positive pressure from dark matter determines the speed and wavelength of a light photon. The source of photons is elements. If an element begins to break apart under negative pressure, it oscillates by converting its constituent solid atoms into soft atoms, continuing until it completely disappears if the interactions persist. This oscillation is so rapid that it creates a photon bombardment sourced from the element. This photon bombardment produces countless (in very large numbers) photons, creating a wave effect.

Describing photons would require an entire book, but don't worry; the mysteries of photons will continue to be unraveled in the hypotheses section and earlier parts of the book.

CHAPTER 6

How Did the Entire Universe, Including Galaxies, Black Holes, Planets, and Stars, Form?

When the Big Bang occurred, countless asymmetrical elements and photons emerged.

Chaos begins, but you must imagine what I'm describing in three dimensions. If you cannot, future documentaries with animations will help you better understand the process. When symmetric dark matter is subjected to high positive pressure, the solid atoms composing the dark matter expand under negative pressure and escape their position. This triggers a chain reaction, and countless soft atoms suddenly appear at the center of space. (This event operates on the same quantum mechanism as a nuclear bomb explosion.) The soft atoms that emerge abruptly form countless clusters, giving rise to photons. (You can imagine photons as air bubbles appearing under the sea.) Large photons disrupt the super symmetric dark matter sequences. The dark matter that enters and remains within the photons without disintegrating accelerates rapidly by sliding through the soft atoms and collides with other dark matter or dark matter sequences. These collisions fragment the dark matter, producing elements. In this chaotic environment, the process I've described repeats countless times. In a space previously containing only dark matter, the explosion at the center has created a vast amount of asymmetrical material (photons and elements). (You may already be aware, but let me remind you: nothing comes from nothing. In some dark matter, solid atoms transform into a soft state to form photons, while the fragmented solid atom clusters from dark matter become asymmetrical elements.)

Because space is so vast, the chaotic environment begins to slow down. (This is because the massive size of space absorbs, or dampens, most of the shockwaves from the explosions, similar to how a nuclear bomb detonated on Earth cannot produce enough chain explosions to destroy the entire planet.)

As the chaotic environment slows, numerous dust and gas clouds (nebulae) form in different regions. These dust and gas clouds occupy external space between symmetric dark matter particles and exert positive force from the inside outward, instantly creating magnetic fields.

As the slowing begins, all the elements and a portion of the photons become trapped within the lattice walls of their own magnetic fields. These magnetic field lattices are then enclosed within even larger magnetic lattices, and simultaneously, all the magnetic lattices begin their journey from the center of the sphere to its surface. Thus, the universe starts to expand. (This is like the journey of an air bubble in the sea rising to the surface.) I'm describing just one, but since space is spherical, magnetic lattices travel in all directions from the center of the sphere to its outer edges. During this journey, the positive pressure from shockwaves caused by supernova explosions within the magnetic lattices, combined with the reflected shockwaves from the initial explosion that were not absorbed, reignites chaos. This marks the beginning of inflation. Yes, a process similar to the inflation theory in mainstream science exists in the Theory of Everything, but the building blocks are different, and it begins shortly after the Big Bang. During inflation, a less intense version of the chaotic environment caused by the initial spark occurs on a much larger scale. Far more elements and photons than those produced during the initial Big Bang emerge, and the amount of material composed of elements and photons increases exponentially.

Billions of large magnetic lattices, each containing countless smaller magnetic lattices, form. Each of these massive magnetic lattices is what we now know as a "galaxy." Thus, you have learned how galaxies form. From now on, when I say "galaxy," remember the massive magnetic lattices containing countless smaller magnetic lattices!

Refresher Paragraph

Everything is determined by a single force. Even natural phenomena labeled as gravity and mass attraction are not fundamental. The primary force in countless events is the resistance that prevents matter from disappearing, preserving its position and volume. This resistance leads to a positive force from the inside outward, making positive force the only fundamental force in nature. Since there is no void at the micro, or quantum, level, nothing can attract anything without contact, and even attraction through contact cannot occur without the use of positive force at its core. Every force, including a vacuum, fundamentally relies on positive force. Please note this and never forget it.

Black Holes

What was happening inside a galaxy during its formation and the ensuing chaos? Let's start with black holes. Galaxies are magnetic lattices, and within each galaxy, there are countless smaller magnetic lattices. What is a magnetic lattice? A magnetic lattice is an invisible wall formed by the symmetric dark matter surrounding an area, compressed from the inside outward, due to the clustering of asymmetrical elements or photons sticking together. The strength of this wall is proportional to the volume, shape, and arrangement of the elements within it. You can see the wall forming the magnetic lattice as an aura when you bring your index finger and thumb about 1 millimeter apart.

I will explain black holes using an analogy: When elements cluster, they accumulate between dark matter particles or their sequences. This accumulation compresses the dark matter in a manner similar to the dome effect. Now, imagine a cluster of clouds in the sky, or if possible, look directly at the clouds. The region with clouds contains a relatively higher density of elements compared to cloudless areas. These dense clusters of elements create massive clouds. These clouds are so large that you cannot see their entirety from the ground; they become visible only from space. You can examine hurricane photos in images of Earth taken from space, which are available online. The area occupied by the elements forming these massive clouds compresses the dark matter. At a certain point, the dark matter becomes so compressed that it opens a hole at the center of the cloud. Because the dark matter in this hole is so dense, the elements forming the cloud cannot easily enter this area. In our analogy, the eye of the hurricane, or this hole, becomes a black hole.

The same mechanical process occurs in galaxies, but it is trillions of times more powerful. A galaxy contains countless smaller magnetic lattices (which include planets and stars, despite being called "small"). The compression of dark matter surrounding these lattices begins outside the first star system, and star systems are separated from each other by a dark matter wall (a magnetic field). The strongest compression occurs at the core of the galaxy. At the galaxy's core, dark matter becomes so compressed that it forms an impenetrable wall while simultaneously creating a vortex. Neither elements nor light can enter this vortex. This area is called a black hole. Black holes form in large galaxies but not in smaller ones. Contrary to popular belief, a black hole does not pull in and destroy everything; it does not take anything in. It acts like an impenetrable wall that reflects everything. Since there are no elements in a black hole, we can now define the dark matter forming the black hole as "dark energy."

There is no stability in the area occupied by a black hole; in other words, within the black hole's region, dark matter flows like water in the form of dark energy. During this flow of dark energy (upward relative to the direction of expansion), even light cannot enter this area. Light coming from other galaxies first shifts toward the weaker edge of the black hole and then accelerates to continue its journey. The scientific community will move from one shock to another; once you learn what you don't know about light, much in science will change.

Star Systems

During the Big Bang and inflation, countless clusters of elements and photons formed. Elements remained scattered like a dust cloud. Photons, being clusters of soft atoms, involuntarily traveled from high-pressure areas to low-pressure areas. Most photons collided with the walls of the magnetic lattice formed by the compression of dark matter in the area occupied by the elements in the dust cloud (similar to the mechanism of radio signals on Earth bouncing off the ionosphere) and dispersed, returning as smaller photons. (Imagine photons as air bubbles under the sea.) These returning photons recombined within the dust cloud, forming larger photon bubbles. If a newly formed photon bubble was too large to withstand the pressure, a supernova occurred, and the chaos cycle restarted within the magnetic lattice. Those that reflected off the lattice and combined to form larger photon clusters began to rotate around each other or other photon bubbles with momentum, merging together. Those that could withstand the pressure after merging became neutron stars, while those that couldn't exploded like a supernova, continuing the cycle. My statement "became a neutron star" might have caught your attention. Neutron stars are photons. Depending on their size and the density of elements around them, they may later evolve into stars. If there aren't enough elements nearby, they remain neutron stars.

Within the dust cloud inside our magnetic lattice, photon bubbles of various sizes formed. Each bubble rotated based on the momentum it had at the moment of formation, creating a vortex in the surrounding dark matter sequences. Now, I'll continue explaining this process using the names from our Solar System, starting with the Sun. Light, or a cluster of soft atoms, travels based on the pressure it experiences. When the pressure is equal from all directions, it temporarily or permanently stabilizes. A massive photon that stabilized within the dust cloud of elements became the core of the Sun, immediately forming its own rotating vortex-like magnetic field and lattice. The Sun's core, through the positive force generated by the area it occupied compressing the dark matter, began to gather and cluster elements from its surroundings into contact with itself. This phenomenon, metaphorically called "gravity," occurred. This process also created a vortex, causing the planet photons around the Sun to orbit it. As the Sun's core formed, eight asymmetrical photon bubbles, along with smaller bubbles around some of them, formed simultaneously. These were the eight planets and their moons. Each had its own volume, vortex, and rotational momentum around its axis. The cores of planets, moons, and stars are photons formed by soft atoms clustering and sticking together between dark matter sequences.

The photons forming the cores of stars and planets, being under pressure from all directions and simultaneously compressed by the dark matter pressing on them, can remain intact for billions of years. However, the cores of moons, being smaller photon bubbles, may fade or disappear over time. Through mass attraction and momentum, each planet and moon core gathered elements around itself based on its volume and rotational momentum. Thus, the Sun, planets, and moons in our Solar System formed. Each large photon created a planet proportional to its volume. Elements without a core nearby formed asymmetrical meteoroids.

With the laws and entities in the Theory of Everything, you can examine each planet in detail and unravel all their mysteries. The internal structure of the Sun and planets begins with a photon core at the center. Around the core, highly compressed elements and dark matter (which may form structures like gold veins) gradually loosen as they extend from the core to the surface. Contrary to common belief, the core and its immediate surroundings in the Sun and planets are cold—or rather, there is no such thing as heat at the core or nearby. As you move from the core to the surface, a proportional loosening occurs, and the temperature rises. Another hypothesis: Lava erupting from a volcano is cold in the deep layers beneath the Earth's crust. The loosening from the Earth's core to the surface is not symmetrical, resembling the thinning of a tree root as it extends deeper. Element arrangements form tectonic plates in stars and planets. For example, the black spots on the Sun are cold regions, with their roots extending from the core in a structure that thins like a tree root from thick to thin.

CHAPTER 7

Fundamental Principles of Micro/Quantum Physics in the Theory of Everything

Quantum refers to a term describing the existence of energy, momentum, and other quantities in physical systems in discrete, specific units.

Quantum is a concept that helps us understand the most fundamental building blocks of nature and how they behave. You may know that "quantum physics goes beyond classical physics, offering new rules and principles that apply in the microscopic world." However, in this book, the quantum world is much simpler. In the Theory of Everything, the initial interaction of everything that occurs in classical physics begins at the quantum or micro level. Since everything is composed of micro and indivisible particles called atoms, it is natural and logical for the first interaction to be triggered by atoms. At the quantum or microscopic level, everything is simpler. Macro events arise from the interactions of countless atoms with each other. Therefore, as we transition from quantum physics to classical physics events, the increasing number of interacting micro particles makes it more difficult and complex to describe events at the micro level. The definitions and events of entities at the micro/quantum level are simple and few in number because the micro level consists of the three classes I mentioned earlier: dark matter, photons, and elements, all of which are clusters of atoms.

For thousands of years, humans were unaware of the micro nature and first tried to understand and define the macro nature. However, knowingly or unknowingly, they have always experimented with dividing a whole into two. Animals do this too. If you break large objects, you get smaller objects; if you combine small objects, you create larger ones.

If you divide a large object in half and keep dividing the resulting pieces in half, the pieces will become so small that your teeth, claws, or fingers will seem enormous compared to the last piece. You can try this with a piece of paper: Take a sheet of newspaper and divide it in half, then keep dividing the pieces in half! Eventually, the divided piece will become so small that your fingers won't be able to hold and divide it because there won't be enough space to grip it from two ends. As you divide an object into smaller particles, the tool used for dividing becomes relatively enormous compared to the last piece. Let's continue dividing with our imagination, using modern technology like a laser to cut. As we keep dividing, we reach the elements. We cut with a laser, which works by sending a photon bombardment. This photon bombardment consists of soft atoms. When soft atoms rapidly collide with an element, the solid atoms forming the element detach and transform into soft atoms. The smallest particle is the atom, which is indivisible. A soft atom cannot divide a solid atom, nor can a solid atom divide a soft atom. Today, some claim they can split an atom! But how? Wasn't the smallest indivisible particle called an atom? If so, what they split shouldn't be called an atom. Initially, the indivisible particle was defined as an atom. Therefore, what they call an "atom" that they split isn't an atom but something else—an element. I believe they split an element but misinterpreted the event, claiming "we split the atom."

In the Theory of Everything, the atom model is simple. First, everything in this theory is composed of indivisible atoms. The reason they are indivisible is this: To divide a soft atom, you need to hold it steady, but that's very difficult. When you apply positive force to a soft atom, it escapes in one direction. (This is also the fundamental cause and effect of entropy.) You can imagine this like trying to split a fly in the air by shooting it with a cannonball. In other words, trying to cut or divide a soft atom with a solid material won't work. Similarly, you cannot split a solid, compressed raw atom with a soft atom. Even if you assume two elements collide at high speed, the solid atoms won't break; due to the collision, they will detach from their position and transform into soft atoms.

In the Theory of Everything, the atom model is an indivisible singular particle. The clusters of solid atoms in this theory are asymmetrical elements. We can equate these elements with the periodic table used today—oxygen, hydrogen, iron, etc. For example, we could redefine the periodic table as "Oxygen is an element composed of x number of solid atoms." In the Theory of Everything, the periodic table consists solely of asymmetrical elements.

Part 1: What Is Energy, How Many Types of Energy Exist, and Where Does the Energy in Our Nature Come From?

Like all other beings, humans have lived intertwined with energy since their existence. Energy is indispensable to living nature. Is energy a natural event or a natural entity? Answer: Energy is the totality of micro and macro events resulting from the structural and positional changes that asymmetrical elements and photons—clusters of atoms forming nature—undergo amidst symmetric dark matter.

I would like to revisit and elaborate on the types of energy according to the Theory of Everything.

Energy is a fundamental physical quantity that expresses a system's capacity to do work. Energy can exist in different forms and transform from one form to another, but the total energy remains conserved. This principle is known as the law of energy conservation.

Types of Energy

1. Kinetic Energy

Kinetic energy is the energy possessed by an object in motion.

Objects formed between symmetric dark matter are influenced by the pressure of this dark matter. At the micro level, photons, which are clusters of soft atoms, are the most affected. The natural source of an object's kinetic energy while in motion is the pressure exerted by dark matter on the object and the object's resistance to this pressure. This pressure naturally accelerates photons to the speed of light and similarly provides natural acceleration to both natural and artificial objects or allows them to maintain their acquired momentum for a period.

2. Potential Energy

Potential energy is the energy an object possesses due to its position or state.

In the Theory of Everything, I defined potential energy as the fundamental force in nature. The only fundamental force in nature is the resistance matter exhibits to preserve its structure, volume, and position due to the impossibility of non-existence; this is simply positive force from the inside outward. Since every particle has an outward positive force, every object possesses potential energy based on its position or state.

3. Thermal Energy (Heat Energy)

Thermal energy is the internal energy of an object and is related to temperature.

It arises from the random motion of molecules. However, in the Theory of Everything, at the quantum atomic level, there are no perceptual concepts like heat, light, cold, or warmth—only the solid and soft states of atoms exist. Thermal energy is the release of soft atoms resulting from the separation of elements, which are clusters of solid atoms. In simple terms, the internal energy of the elements forming an object creates the perception of heat as solid atoms transform into soft atoms.

4. Chemical Energy

Chemical energy is the energy stored within chemical bonds and is released during combustion or chemical reactions.

In the Theory of Everything, elements can form bonds (by sticking together) between dark matter. The separation of these bonds—meaning the detachment of the adhered elements (which may also involve the event described under thermal energy)—is a chemical reaction.

5. Electrical Energy

Electrical energy arises from the movement of electric charges and is used in electrical circuits.

In the Theory of Everything, electricity and its energy result from the flow of electrons between elements. I will explain this in detail in the electricity section later, but for now, let me offer a surprising proposition: Lightning and thunderbolts are not electricity. Storing electricity from lightning or thunderbolts is inefficient, if not impossible.

6. Nuclear Energy

Nuclear energy is the energy stored within an atom's nucleus and is released during nuclear fission and fusion processes.

However, in the Theory of Everything, I will make a correction: Since atoms are indivisible, they do not have a nucleus. Nuclear energy is the pure clustering of elements between dark matter sequences. If this clustering contains 100% elements, it becomes a nuclear bomb; if it contains 90% elements, it is used as a nuclear fuel rod.

7. Light Energy (Electromagnetic Energy)

Light energy is the energy carried by electromagnetic waves and can take various forms, such as sunlight, radio waves, and X-rays.

According to the Theory of Everything, light energy is a cluster of soft atoms. Since clusters of soft atoms are asymmetrical, they exist in numerous forms, such as X-rays and radio waves.

The Law of Energy Conservation

In my book The Theory of Everything, I have addressed many gaps in mainstream physics and corrected misconceptions. However, I sometimes feel the need to teach fundamental concepts of physics to readers unfamiliar with the subject, and energy is one of them. I liken energy to the concept of time: In short, the faster a material moves, the more energy it produces. In other words, energy is a concept that emerges relative to the changes and speed of matter.

The law of energy conservation states that the total energy in a closed system remains constant. This means energy is neither lost nor created from nothing; it only transforms from one form to another. The Theory of Everything supports the conservation of energy, but there is no such thing as a completely closed system in space, including space itself. Since everything is composed of indivisible particles—atoms—completely closed systems are rare. Could black holes be considered a closed system? If the universe didn't expand and galaxies remained stationary, we could speak of a closed system, but that's not the case. Neutron stars and the cores of planets or stars, while completely closed systems, have the potential to explode or disperse. In conclusion, in a closed system, the total number of atoms remains constant, and thus the potential energy of the atoms remains constant as well.

For example, in the motion of a pendulum, when the pendulum reaches its highest point, its potential energy is at a maximum, and its kinetic energy is at a minimum. At its lowest point, the potential energy is at a minimum, and the kinetic energy is at a maximum. Throughout this transformation, the total energy remains constant.

Energy Transformation

Energy can transform from one form to another in various ways. Here are a few examples:

1. Mechanical Energy - Thermal Energy

Due to friction, a moving object converts its kinetic energy into thermal energy. Friction is caused by applying positive force. Since positive force simultaneously creates negative force in some quantum regions, elements release soft atoms, generating thermal energy.

2. Chemical Energy - Electrical Energy

In batteries or cells, chemical energy is converted into electrical energy. Batteries and cells store soft atoms (this is one of the cases where small photon clusters remain fixed in the form of electrons).

3. Nuclear Energy - Thermal Energy

In nuclear reactors, nuclear energy released from core reactions is converted into thermal energy. I will address this topic later.

4. Solar Energy - Electrical Energy

Solar panels convert sunlight (electromagnetic energy) into electrical energy. This is the trapping of a portion of soft atoms from the photon clusters coming from the Sun.

Conclusion

Energy is a fundamental concept that defines the capacity to do work and exists in many different forms. The transformation and conservation of energy are fundamental principles of physics, playing a critical role across a wide range of applications, from daily life to industry, transportation, and electronic devices. These foundational insights about energy are a key part of modern scientific and technological understanding. I have aimed to support these concepts with the Theory of Everything, although I have also made some propositions that may seem contrary to mainstream views—but don't worry, keep reading!

Part 2: States of Matter

The atoms that form the building blocks of everything in nature and our environment were not created from nothing; they have only transformed by changing their position or volume.

The phrase "Everything is one, and one is everything" beautifully describes the atom.

As you may recall from my previous explanations, an atom has only two extreme states, and these states apply solely to the atom, the smallest indivisible unit of matter. The extreme states of an atom are its soft and flexible state, where its volume is larger, and its raw, solid, and unexpanded state, where its volume is smaller. It's acceptable to metaphorically refer to the solid and soft states as hot and cold, but we must not forget that in nature, hot and cold are perceptions arising from the vibration or motion speed of matter.

Over 99% of space is covered with symmetric and super symmetric sequences of dark matter and energy (dark matter exists in areas dense with elements, while dark energy exists in areas where elements are absent).

The most well-known states of matter are solid, liquid, and gas. These primary states have countless levels, much like the countless levels between the solid and soft extreme states of an atom.

I will explain the states of matter from the quantum level to the scale we perceive, starting by introducing magnetic fields, which are particularly strong in solids, in a way that aligns with the Theory of Everything.

Magnetic Field

Magnetic fields play a crucial role in our nature. From the states of matter to electricity, magnetism to microgravity, and microgravity to colors, magnetic fields, along with their levels such as aura and magnetic lattice, have a significant impact.

Let's conduct a thought experiment with you, the readers; physicists who wish to can also perform this as a physical experiment. The name of our experiment is "Understanding and Recognizing Dark Matter," or in short, KARanlık MAddeyi Tanıma ve ANlama, which we'll call "KARMATAN" (the capital letters represent the acronym in Turkish).

Materials Needed for the Karmatan Experiment:

a) A 2-square-meter empty pool with all four sides enclosed.

b) Enough hexagonal, flexible materials (like rubber or similar, with minimal elasticity), preferably 1 or 5 cm in diameter, to fill the pool's area (the goal is to create a honeycomb-like structure).

c) A rope to create an imaginary projection line for observation.

d) Solid and flexible asymmetrical materials, smaller than the hexagons.

These are the materials needed for our experiment to understand magnetic fields. Let the pool represent space, the symmetric hexagons represent dark matter, the asymmetrical solid materials represent elements, and the soft ones represent photons. The rope is solely for observation. In our experiment, arrange the hexagonal materials in the pool like a honeycomb, filling it entirely. Then, using the rope, create projection lines 10 centimeters inward from the edges at intervals, ensuring they don't touch the materials from above. Making a square that leaves a 1-square-meter empty space is sufficient. Finally, forcefully insert the asymmetrical materials into the gaps between the hexagons within the 1-square-meter area enclosed by the rope in the center of the pool.

What happens? By creating external space between the dark matter in the center of the pool and placing elements and photons there, we cause the hexagonal dark matter in the pool to compress even further. This increases the positive force pressure from the inside outward with the elements we compressed in the center. As a result, the dark matter compresses more toward the pool's edges with a proportional force. The areas aligned with the ropes become the magnetic field of the object formed by the elements. The first rope alignment at the center, being the most compressed area, becomes the object's aura and lattice. (This can be likened to packing a suitcase: as you add items to the center, the clothes at the edges get more compressed. In other words, the more pressure an element cluster exerts from the inside outward, the more the surrounding dark matter compresses horizontally.) Now imagine this in three dimensions and in space. Magnetic fields are particularly strong at the surface of an object. A magnetic field also acts as a lattice that holds objects together, serving as an invisible wall that protects the object from external influences. You can see a magnetic field by bringing your index finger and thumb 1 millimeter apart and looking at the light. Due to this magnetic field, you cannot directly touch something without applying sufficient force. It is the magnetic field that protects our Earth from the Sun's harmful photons. Since the magnetic field also acts as a magnetic lattice, it significantly influences the phenomenon known as gravity. Magnetic fields are composed of dark matter.

One of the key factors behind the countless levels of states of matter is the magnetic field. The arrangement, size, and shape of elements among dark matter determine the magnetic field, leading to materials with countless levels of solidity, such as iron, feathers, human skin, or ice.

I hope this thought experiment has helped you better understand magnetic fields. The solidity level and the resulting magnetic field in the states of matter vary depending on the arrangement, structure, and density of elements among symmetric dark matter.

In the liquid state of matter, elements are more dynamic. Liquids also have magnetic fields, but their magnetic lattice—or the magnetic wall formed outside the element cluster—is weaker. This weakness allows liquids to easily separate into cells. For example, while the ocean as a whole has a strong magnetic lattice, a wave collision can open the magnetic lattice wall, causing the liquid to separate into cells like water droplets. The primary reason for the liquid state of matter is usually the small size of the elements forming the liquid. Liquids are generally composed of many small asymmetrical elements, such as oxygen or hydrogen.

The gaseous state of matter similarly forms with elements among dark matter, but the element density is very low.

Since every object is made of indivisible atoms, it can transform from solid to liquid to gas, or vice versa, when the necessary artificial or natural conditions are met. For example, how does water transition from a liquid to a solid state? Cooling occurs when the speed of motion decreases, while heating occurs when the speed increases. The factor determining an object's motion speed is dark matter. In this context, when dark matter compresses the elements between them with positive force, the elements' motion is restricted, and cooling begins. Conversely, if the space between dark matter loosens, the elements' motion speed increases, and heating begins. The pressure force that dark matter exerts on each other is interconnected; if there is a cold area in one place, a hot area must form elsewhere. This applies to all planets and stars.

In the diagram below, you can see the gaseous state of matter.

Among symmetric dark matter, there are randomly arranged asymmetrical elements of varying sizes and shapes. In the gaseous state, elements are sparse.

The sparsity of elements allows photons to pass easily through the dark matter. Additionally, this sparsity keeps the pressure dark matter exerts on each other at a low level. Since the gaseous state of matter forms with sparse and small elements, the magnetic lattice wall of the area containing gas is weaker than that of liquids. The magnetic lattice wall of a gas forms in very large areas, like the ocean example I gave for liquids. For instance, the magnetic field of the gas we call "air" covering Earth's sky begins after the troposphere. Beyond the troposphere, elements become sparser, and dark matter exerts stronger pressure on each other. The magnetic lattice wall makes it difficult for elements to escape into space and also prevents some photons from escaping. Some of the photons launched from the Earth's surface toward space bounce back from the magnetic lattice wall. This phenomenon enabled the discovery of radio broadcasts years ago.

Gas state of matter

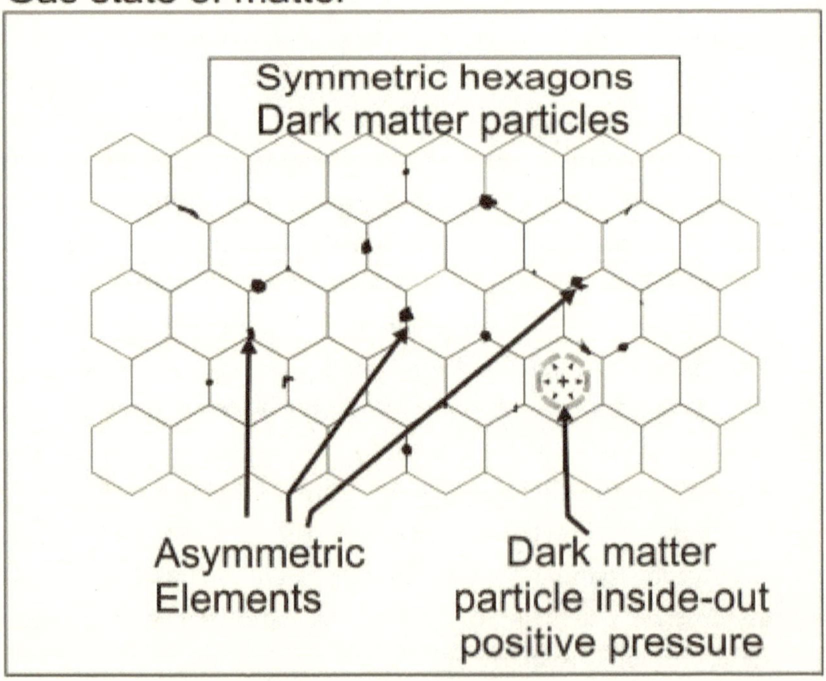

The three states of matter are determined by the density, size, shape, and arrangement of elements among dark matter particles. As we saw earlier in the gaseous state of matter, elements are randomly and sparsely distributed among dark matter. As elements become denser, their visibility increases; thus, as matter transitions from gas to liquid, it eventually forms an object. In other words, an observable change is inevitable as elements become denser.

Let me explain: When elements are very sparse, the likelihood of a photon colliding with an element is low. However, as the number of elements increases, the probability of a photon colliding with an element rises. For example, when elements in the sky are very sparse, the sky appears clear, and all stars and sunlight reach us directly. As elements in the sky become denser, the invisible gas we call "air" forms visible clouds in a gaseous state in certain areas. When elements become even denser, raindrops form within the clouds. If the elements densify further while dark matter simultaneously compresses, the raindrops turn into snow or hail. In conclusion, the density and arrangement of elements among dark matter not only enable the formation of all objects but also determine the states of matter. Both the gaseous and liquid states of matter have magnetic fields, but these fields are weak. The magnetic field of gas and liquid states only becomes stronger when it spans a very large area.

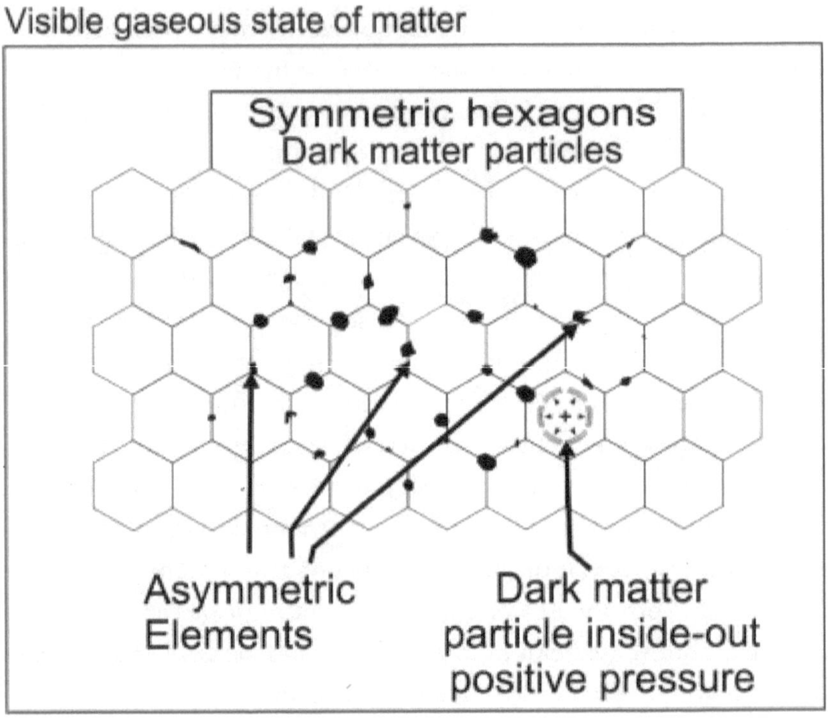

Dark matter is in a constant state of action and reaction with elements and photons. Elements and photons exert an outward positive force on dark matter, while dark matter resists this effect by applying an inward positive force in response, or the reverse occurs. In essence, the necessity of existence results in a continuous "arm-wrestling" or "force dance" between photons, elements, and dark matter. On a larger scale, there is a force competition between the regions formed by dark energy and dark matter, which is the fundamental reason for the universe's expansion.

Dark matter most easily wins this force competition against photons. The intense pressure from dark matter causes photons to slide very quickly. Since the magnetic wall of matter in its gaseous state is weak, external photon bombardment limits the gas's ability to remain dense. This is why matter in its gaseous state appears visible or invisible for certain periods.

The liquid state of matter typically forms most easily through the transition from gas to liquid. When elements among dark matter are in a gaseous form, the transition to liquid occurs if external elements are introduced or if the gas becomes denser in a smaller area. In the model below, I have attempted to illustrate the liquid state of matter.

However, a detail arises here: In the liquid state of matter, dark matter detaches from its sequences. This detachment facilitates the fluidity of liquids, especially water. In the model's similar element arrangements, if the elements are composed of very small particles, the fluidity is high; if they are made of larger particles, the fluidity is slower. Additionally, even if composed of small elements, if the solid atoms on the surface of dark matter and elements wear down, resulting in a solid base and a soft tip, a superfluid liquid forms. An example of a superfluid liquid is the blood in our veins. We can explain the structure of matter in its liquid and fluid state using the analogy of a ball bearing: Let's assume the central ring of the bearing represents dark matter detached from its sequence due to small elements and electrons; the bearing's balls represent light and small elements; the space between the balls (electrons) represents soft atom particles; and the outer part of the bearing represents the boundary in contact with other dark matter. Thus, with the positive pressure each particle experiences from the outside inward and the resistance it exerts from the inside outward, a force dance begins, creating the dynamic environment of the liquid. When the forces balance, the liquid behaves like a still lake.

Liquid state of matter

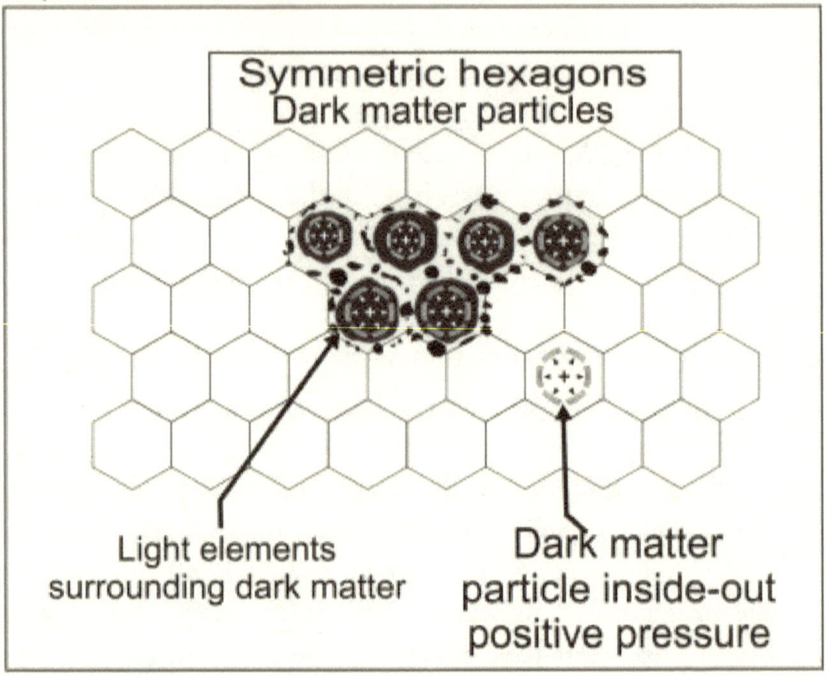

What happens if the positive force pressure exerted by dark matter on each other and on elements in a given area increases? In this case, the situation shown in the diagram below occurs. This process is similar to water turning into ice, meaning a transition from liquid to solid takes place. On Earth, this phenomenon is more commonly observed during the winter season. The transition from liquid to solid can also occur with the introduction of dense and large elements from the outside. For example, adding flour to water makes the liquid less fluid, increasing the internal pressure in that area. This increased internal pressure gradually pushes some of the lighter elements forming the liquid out of the area. When you heat dough, the influx of photons causes the elements to expand, accelerating the process, and the dough dries, transitioning into a solid state.

The solid (ice) form of water composed of elements

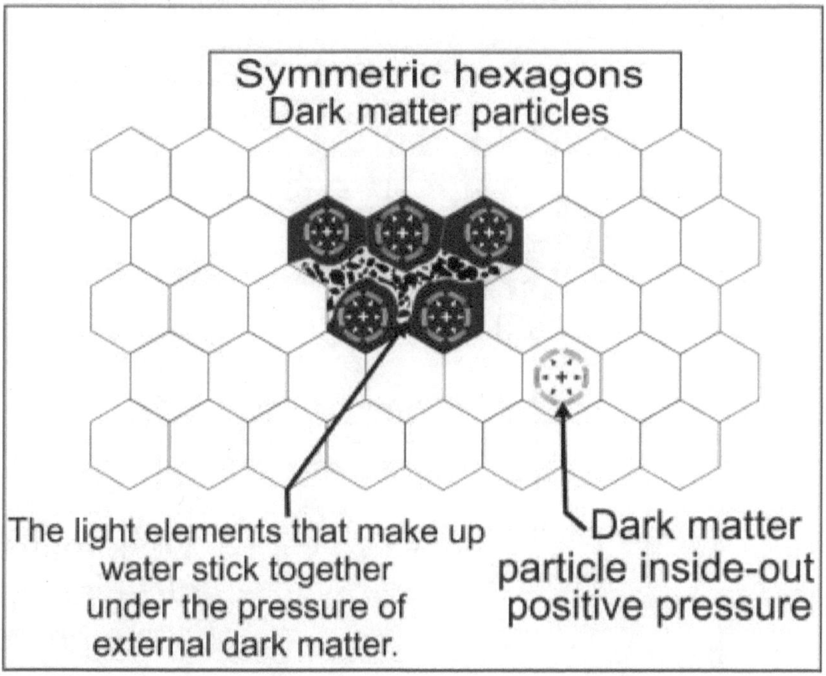

As seen above, the positive force pressure exerted by dark matter from the outside inward compresses the elements, causing them to stick together and form larger elements. This is how water turns into ice.

The arrangement of elements into larger particles constitutes the solid state of matter. Natural solid states, such as iron, stone, steel, tree trunks, and rocks, contain larger but fewer elements compared to water. The molecular bonds formed by these larger elements can create countless combinations depending on their asymmetrical shapes and sizes, such as wood, plastic, iron, gold, silver, and so on.

Solid state of matter

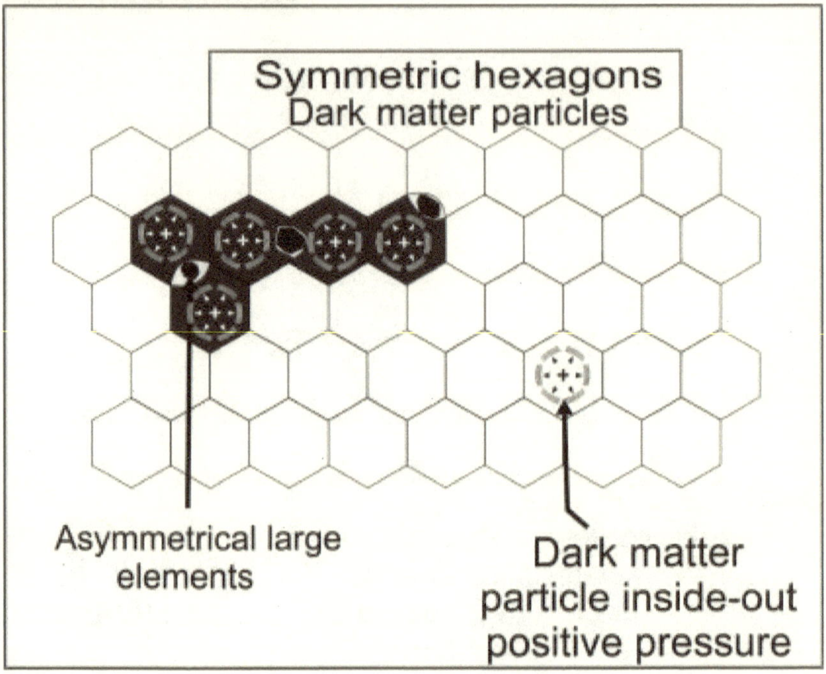

Part 3: What Are Magnetism, Magnetic Effects, Magnetic Lattices, and Fields? How Do They Form and Function?

The first time I encountered a magnet in my life, I was quite surprised. The effects of the magnet didn't inspire awe in me but rather curiosity. When I brought it close to a small piece of metal, the magnet attracted the metal; when I brought it near a larger piece of metal, the metal attracted the magnet. A natural phenomenon was occurring, and like most people, I interpreted and described it as "It attracts." I liked the magnet I had, so I kept it instead of throwing it away. Later, when I found another magnet, I noticed that one side of the magnet attracted while the other repelled. Thinking I had satisfied my curiosity with my own definition, I set the magnets aside and returned to my daily life.

However, after discovering the Theory of Everything, I realized I had completely misinterpreted everything about magnetism. The conclusions derived from inductive reasoning showed that nothing in nature could occur without positive force.

I can briefly define the cause of magnetism as the concentration of dark matter around a cluster of elements. Recall the Karmatan experiment: When we inserted external materials into the center of a pool made of symmetric, hexagonal, and flexible materials, the area occupied by those materials compressed the hexagons proportionally from the inside outward. This compression of dark matter led to the formation of the magnetic field and lattice. The magnetic field lattice wall is like snow accumulated on a tree after a winter snowfall or frosting on a cake. We cannot directly see this magnetic field lattice wall, but we can observe a thin membrane on the object through the refraction of light. This is because photons carrying visual information travel by sliding between dark matter particles.

Therefore, in areas where dark matter is densely packed, light bends, and the photon cluster loses elements.

When looking at a candle flame while bringing the tips of your index finger and thumb very close without touching, you can see the light refracting in that area and a membrane forming on the surface of your fingertip. The light refracts because of the magnetic lattice wall formed by dark matter when we bring our fingers close together. Magnetic fields can span an area several times larger than the object itself. The size and strength of the magnetic field depend on the arrangement and size of the elements forming the object. For example, the magnetic field of a plastic ball is not the same in size or strength as that of a metal ball; their magnetic lattice walls differ in durability.

Dark energy particles in two dimensions are symmetrical in the shape of a hexagon.

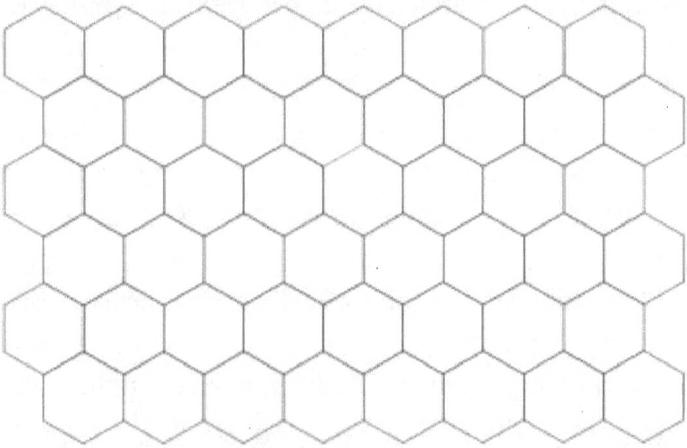

Since there are no elements between the particles, the dark energy arrays are symmetrical.

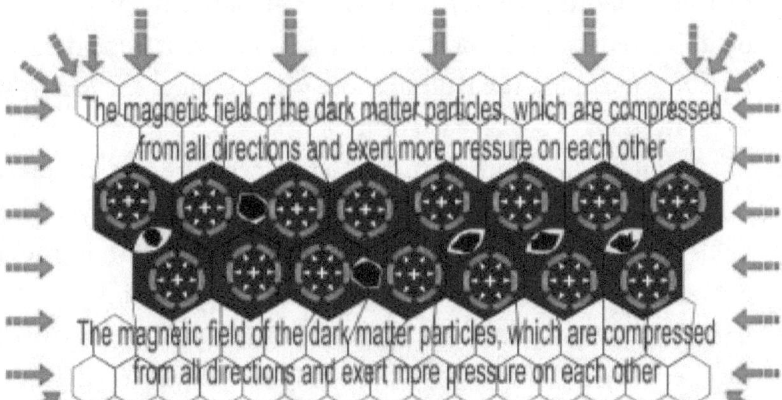

Dark matter always presses the elements from the outside in. The elements between the dark matter particles always exert force from the inside out, resisting the external force they are exposed to.

In the upper diagram above, you can see symmetric dark energy particles in the shape of hexagonal-cubic grains. (We previously stated that if there are elements between the symmetric structures forming space, it is called dark matter; if there are no elements, it is called dark energy.)

Recall the Karmatan experiment, but this time imagine it in 3D: Let's assume we trap dark energy particles in an imaginary space, say a 10 cm^3 area. Suppose there are 1,000 dark energy particles in this 10 cm^3 space. In this imaginary area, each particle exerts equal pressure on the others, making dark energy particles super symmetric. If we sprinkle external elements among these particles, similar to what is shown in the diagram on the right, we add an extra component to the 10 cm^3 imaginary space. The moment elements are introduced, dark energy transforms into dark matter, and depending on the density of the elements, an object in solid, liquid, or gaseous state forms, along with a magnetic field around it. (Think of this as adding apples to a basket already full of balloons: the balloons represent elements, and the apples represent dark matter. Due to the apples, the balloons lose their original shape and stretch to fill the gaps between each other.)

Magnetic fields are a crucial factor in nature and are present in everything large enough to be visible to the naked eye. Magnetic fields prevent natural entities from dispersing from the inside outward while also protecting them, to a certain extent, from harmful external elements and photons. On a macro scale, the magnetic field holds our planet together. Thanks to the magnetic field, the photon bombardment from the Sun does not harm living organisms, except in extreme conditions. Overcoming the magnetic field from the inside outward is slightly more difficult than from the outside inward, which is why rockets struggle to leave Earth. For this reason, there is a magnetic barrier that grows stronger from Earth to the Sun, the Solar System, and the galaxy, increasing in scale.

Each magnetic field acts as a lattice for the element cluster at its center. Due to these magnetic lattices, a celestial body in star system A cannot cross the space between stars to reach star system B. This applies to our galaxy as well: a star in our galaxy must overcome the galaxy's magnetic field and wall to leave it. Magnetic fields also play a role in phenomena known as mass attraction and gravity.

Elements come in countless different shapes, masses, and volumes, allowing them to arrange in numerous combinations among dark matter particles. This is why element clusters create diversity in nature, and a magnet is one of them. The magnetic field of a magnet, due to the arrangement and shape of the elements within the cluster, metaphorically "attracts" metals.

If every element cluster is surrounded by a magnetic field, then it is inevitable that magnetic fields interact before the element clusters themselves do. You can observe this with the naked eye right now: Bring your index finger and thumb about 1 millimeter apart and look at a bright object through that 1-millimeter gap! You'll notice the initial boundary of your magnetic field like a membrane, and the membrane outside your skin merging into each other. The effects of the magnetic field are more pronounced in magnets. In magnets, magnetic fields are much stronger compared to organisms like us because dark matter compresses each other more intensely.

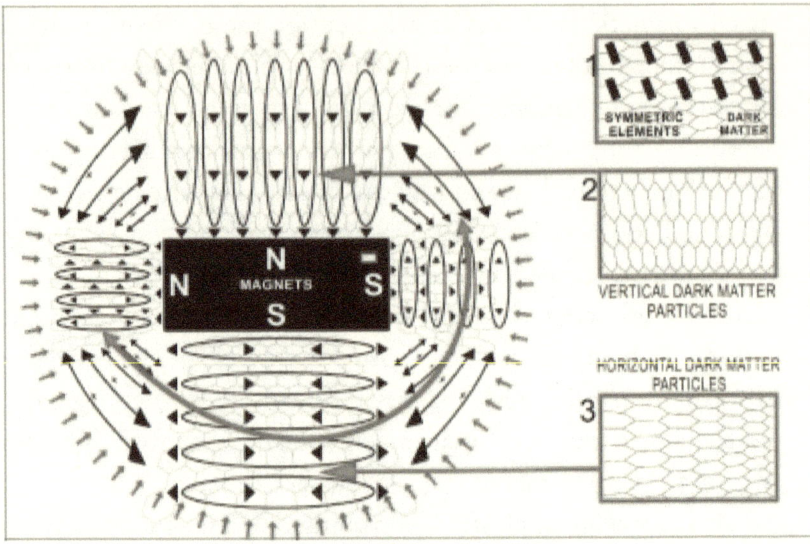

In the diagram above, we consider a rectangular magnet, where (N) represents the north pole and (S) the south pole. (The poles of a magnet have been defined for navigation purposes using a compass. In reality, the poles of a magnet are determined by the arrangement of the elements forming the magnet. In compasses, the needle moves relative to the density of dark matter, which is more compressed at the poles.)

The diagram shows the pressure of dark matter (indicated by the outermost red arrows) outside the magnet cluster. In the upper right, the arrangement of elements among dark matter particles is depicted.

All metals and stones, including natural magnets, are composed of heavy, or large, elements. These large elements are less numerous in nature compared to smaller ones. For this reason, the strong attraction or repulsion reactions, like those of a magnet, are not commonly observed among all objects on Earth. An object made of large elements is typically in a solid state, due to the positive force exerted by dark matter from the inside outward and by the dark matter outside the cluster from the outside inward. To better understand this, let's conduct a thought experiment by imagining the

Karmatan experiment setup in 3D.

For example, in a 10 cm^3 space, let there be 100 dark matter particles and 100 air elements made of small, sparse elements distributed sparsely within this space. Since the elements are small, the 10 cm^3 space will not expand significantly; however, let's assume the number of dark matter particles in the space decreases to 99. Now, let's fill another 10 cm^3 space, originally containing 100 dark matter particles, with 100 iron elements that are half the size of the dark matter particles. In this 10 cm^3 space, there will be 50 dark matter particles corresponding to the 100 iron elements. The iron elements, entering the imaginary space from outside and occupying space, push half of the dark matter particles outward, compressing them. This results in the iron cube having a stronger and thicker magnetic field. In conclusion, the number of dark matter particles in the space decreases relative to the area occupied by the elements; however, dark matter does not disappear—it is merely compressed just outside the space.

The elements in the internal structure of a magnet are symmetric and larger compared to, for example, the elements forming organic beings. This is why the magnetic field on the surface of a magnet is dense and thick. The molecular arrangement of a magnet resembles the diagram in the upper right. We can better understand this by imagining the dark matter in the center of the Karmatan experiment as hexagonal and equipped with a spool-like twisting mechanism. When large elements form a molecular arrangement due to environmental factors, they twist the dark matter like a spool, disrupting the symmetry and shape of the dark matter in its sequence.

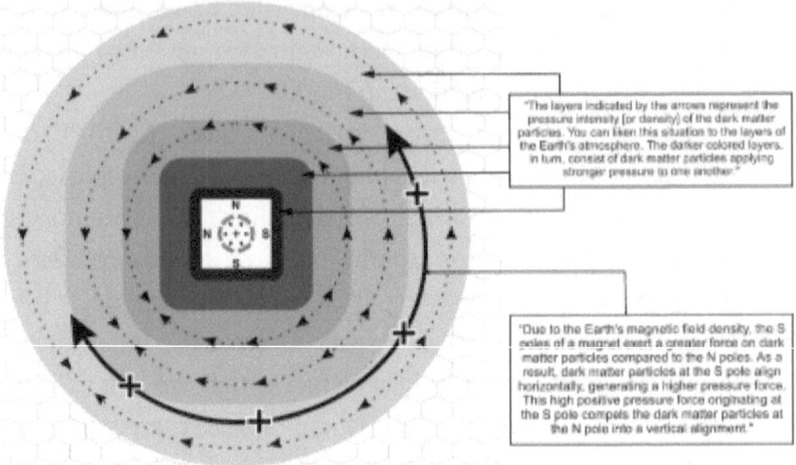

When the symmetric arrangement of dark matter is deformed due to the arrangement of elements, force directions as shown in the diagram emerge among the dark matter particles forming the magnetic field. It's challenging to explain these phenomena with 2D diagrams, but it might be easier if I put it this way: "The space formed by dark matter bends with the presence of elements." All magnetic fields are created by the bending of the dark matter space. The magnetic field of a magnet is also a form of space bending, but the shape of this bending is a rare type (space here refers to dark matter).

In the diagram, I've labeled the left and upper left parts as (N) north, and the right and lower right parts as (S) south. The gradient from dark to light outside the white square represents the magnetic field layers, transitioning from strong to weak. In other words, the dark areas indicate compressed dark matter particles, while the lighter areas show progressively looser dark matter. The property of magnetism arises from the molecular arrangement proportional to the size of symmetric elements, occurring during the object's initial formation. We call such objects natural magnets (magnetite). Ultimately, the prerequisite for any object to exhibit magnetic properties lies in the size, shape, and molecular arrangement of the elements forming the object, influenced by environmental factors during its formation. For instance, if we hammer a piece of iron, we affect the molecular arrangement within the cluster, and with each strike, we compress the elements—and thus the dark matter—further, potentially turning the iron into a temporary magnet. (It's temporary because the dark matter, compressed by external intervention, resists from the inside outward and gradually returns to its original state over time.)

In the first diagram below, you see a square-shaped magnet at the center. The dotted lines (vectors) represent the total extent of the magnetic field. At the center of this field, the space occupied by the elements forming the object has been taken from the space of dark matter particles. Consequently, the overall magnetic field is formed by the stretching of dark matter particles in the (N) direction and their compression toward the (S) directions within the object. The compression of dark matter in the (S) direction, with the resulting high positive force pressure, further flattens the dark matter in the (N) direction (indicated by the crescent-shaped half-circle arrow). In other words, the (S) directions suppress the (N) direction.

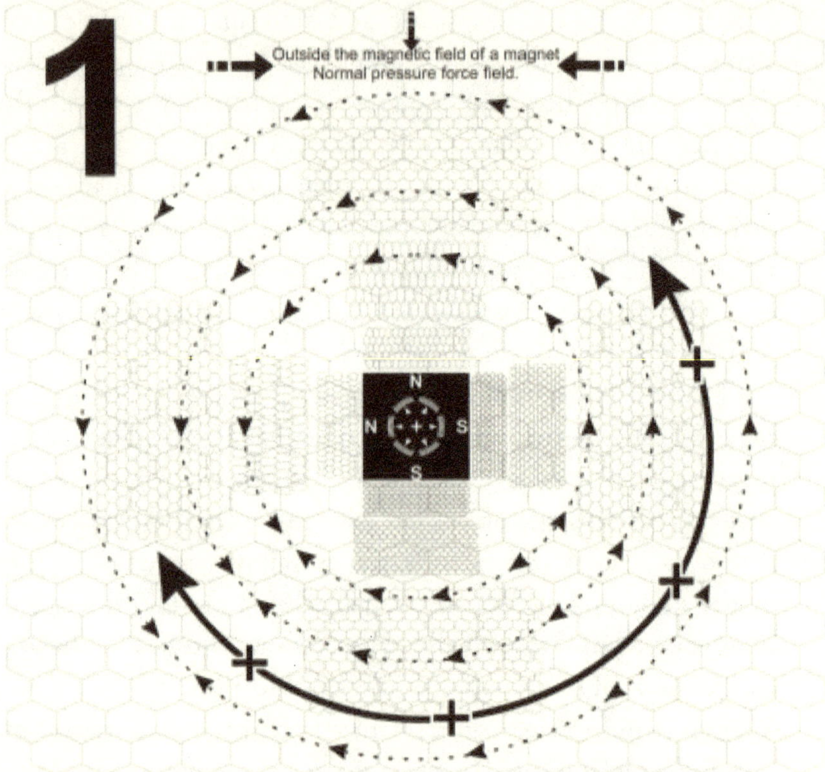

In the second diagram below, there is a magnet at the center with a metal iron piece above it. The magnetic field of the metal iron, due to its proportional molecular arrangement, covers a smaller area. When the metal iron enters the magnet's magnetic field, it is subjected to the magnet's magnetic field pressure in addition to its own. As a result, the total dark matter pressure it experiences will be greater than the pressure outside the magnet's magnetic field boundary. Thus, the magnet's magnetic field, combined with the pressure of the surrounding dark matter particles, pushes the metal iron toward the magnet with a force that increases proportionally with the distance. This phenomenon is not a vacuum or negative pressure; only positive forces are at play here. When the metal iron sticks to the magnet, the magnetic field lattice walls—where the most compressed dark matter resides—merge, making it difficult to separate the magnet. (Note that in the areas where they stick, normal matter atoms do not actually touch each other.)

2

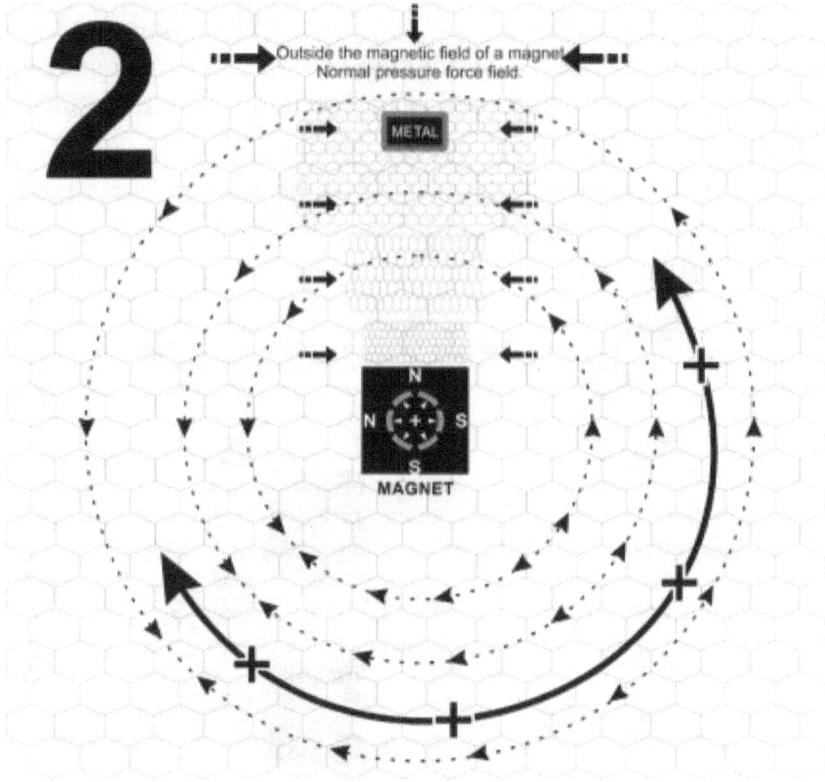

In the third diagram below, we consider two magnets of the same size. Since the magnet's formation, the dark matter in the (N) direction has been flat and flexible. When you rotate the magnet, these stretched dark matter particles rotate as well. If you bring the same poles of two magnets closer together, the pressure from the (S) directions of both magnets doubles between them. In other words, the stretched and flattened dark matter particles are now subjected to twice the pressure, making it difficult for the magnets to approach each other from the same poles. This is solely due to the positive pressure force of the dark matter from the (S) poles. You can imagine this like trying to cross a river with a strong current: Picture yourself and your friend as magnets, with you on one bank and your friend on the opposite bank. The strong current of the river represents the dark matter pressure, making it hard to cross to the other side. However, if the magnets face opposite directions, the dark matter pressure from the (S) direction is supplemented by the pressure from outside the cluster, pushing the entire area together and allowing the magnets to stick easily.

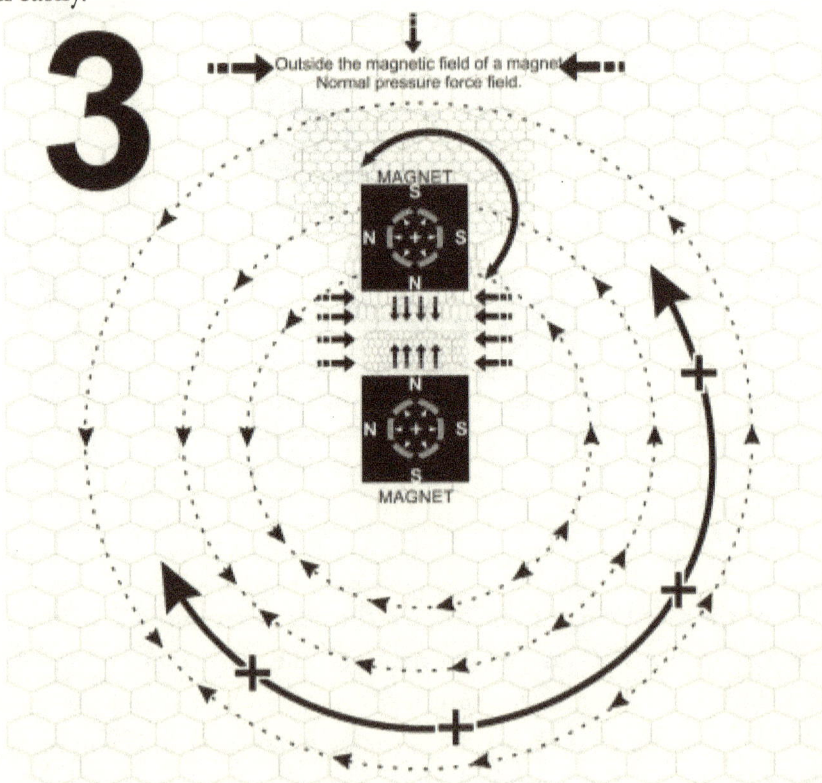

Part 4: Photon

We can only observe everything that physically constitutes our nature in the form of clusters. The closer we look at a cluster of scattered or adjacent particles, the more clearly we see micro particles; the farther we look, the more we perceive it as macro particles. For example, when we look at stars, galaxies, or planets from a distance, we see them as a single particle, but as we get closer, we realize they contain many particles. The reason for all this is photons. Moreover, everything we look at appears two-dimensional as the distance increases and three-dimensional as the distance decreases, which is also due to the events photons undergo during their journey.

Photons are shapeless clusters of soft atoms with light mass that carry the visual information of micro particle clusters. A photon is a cluster of atoms in a soft state. Therefore, according to the proposition "We can only observe everything that physically constitutes our nature in the form of clusters," we can only observe photons when they are in clusters (as light). No matter what we look at, we don't perceive the object itself but rather the photon clusters that are either reflected from it or have directly detached from it to become photons. This is because photons are also composed of particles with mass and volume, but they are much lighter relative to the space they occupy. Consequently, when photons reach our eyes or devices, their potential and kinetic energies are discharged by interacting with the elements they contact.

As you may recall, we referred to the smallest indivisible and unbreakable unit of matter as the "Indivisible Singular Organic Matter," or simply "ATOM." An atom has two extreme states: a small, solid state by volume, and a large, soft state when stretched. While solid atoms can stretch to become soft, soft atoms cannot compress back into a solid state. This is due to the impossibility of creating an environment—naturally or artificially—that can compress a soft atom from all directions. For instance, we can push toothpaste back into its tube by applying pressure from all sides, but we cannot build a device to compress a soft atom from all directions. This is because a soft atom is so small that it has many places to escape when compressed. Naturally, when subjected to dark matter pressure, it slips away.

Solid and soft atoms are indivisible particles. Considering that devices are also made of elements formed by countless atoms coming together, it is impossible to observe a single atom or enable devices to detect it. If we cannot create such devices, everything we can see will either originate from clusters of elements made of solid atoms or be photons—clusters of soft atoms—reflected from those clusters or originating from another source.

Photon: The Quantum Particles of Light

Photons, one of the most fascinating and mysterious entities in the world of physics, are commonly known as "the fundamental building blocks of light and electromagnetic radiation," but they possess far more properties than generally assumed. Photons are quantum particles that cannot be directly observed at the micro level, yet their existence and behavior help us understand the structure and functioning of the universe. Without them, explaining the universe's most fundamental interactions would be impossible. Photons play a critical role in all forms of communication, whether natural or technology-assisted.

The concept of the photon emerged in the early 20th century with the development of quantum mechanics. In his famous 1905 paper, Albert Einstein proposed that light travels in discrete energy packets, introducing the existence of photons and laying the foundation for the wave-particle duality concept, which states that light exhibits both wave and particle properties.

Photons represent the quanta of electromagnetic waves formed by the continuous transformation of electric and magnetic fields into each other. These small yet energetic particles move at the speed of light due to dark matter pressure and have a light mass. Their soft and lightweight structure allows them to carry potential energy, which transforms into increased momentum when accelerated at the speed of light, enabling them to interact with matter and trigger various physical phenomena.

Understanding the nature of photons is crucial not only for theoretical physics but also for technological innovations. Modern telecommunication systems, lasers, photovoltaic cells, and many optical devices operate by leveraging the properties and interactions of photons. Furthermore, advanced fields like quantum computing and quantum cryptography aim to revolutionize information processing and security by harnessing the quantum properties of photons.

In this section of the Theory of Everything, we will explore the fundamental properties of photons, their roles, and their interactions in nature in detail. Welcome to the astonishing and captivating world of photons!

Photon clusters are composed of soft atoms. Although soft atoms can occasionally form directly from dark matter in small amounts during violent phenomena like the Big Bang, supernovae, nuclear explosions, lightning, or thunder, they typically originate from the solid atoms within elements. From the sunlight to the flame of a candle, nearly all light and heat you've encountered in your life stem from the release of solid atoms forming elements, which then create clusters of soft atoms.

Asymmetrical photon clusters can give rise to the entire spectrum of light—including gamma rays, X-rays, ultraviolet, visible light, infrared, microwaves, radio waves, and unknown types—as well as countless natural phenomena and elements like lightning, thunder, electricity, and the cores of planets and stars. Since photon clusters are made of soft atoms, they can be classified based on the number of soft atoms in the cluster. While we cannot precisely determine the exact number of soft atoms in a photon, assigning numbers proportional to the frequency of light ensures that mathematical calculations remain accurate. I can suggest a method for this: For example, by initiating decay or emission in the lightest element, hydrogen, under negative pressure, we can calculate the duration and the number of electron/photon waves produced. The assigned number reveals how many solid atoms hydrogen consists of, and the frequency of the emitted light indicates the number of soft atoms in the photon cluster, allowing for the classification of elements and photons.

All elements and photons in nature exist between symmetric dark matter sequences. Photons are clusters of soft atoms formed when solid atoms stretch and expand in volume. Due to their soft and shapeless structure, photons behave like a liquid under the pressure they experience between dark matter, making photons and their other form, electrons, the most superfluid entities in nature. The difference in fluidity between an element and a photon under dark matter pressure is akin to squeezing a stone versus water with your hand. You've witnessed this countless times in your life: lighter objects are affected by forces more quickly and easily compared to heavier ones, and this principle holds true at the micro level as well.

A photon is a superfluid that rapidly slides between dark matter. Moving at the speed of light, its potential and kinetic energies—despite its light mass—can give rise to powerful phenomena, one of which is waves. Photons not only travel at the speed of light but also rapidly detach from their source, forming clusters. As soon as a photon forms, its escape from the element it originated from creates a photon bombardment. This bombardment, occurring in directions determined by environmental factors, causes photon particles to form a wave function. Simply put, a photon is always a cluster of particles, but its movement at the speed of light and the separation of millions of photons per second from its source allow us to perceive photons as waves as well.

In the Theory of Everything, the atom model is quite simple. The smallest indivisible solid particle is the atom. I have defined the process of a solid atom stretching and expanding into a soft state as a "soft atom." Asymmetrical clusters of solid atoms are called "elements," clusters of soft atoms are called "photons," and the largest cluster of symmetric solid atoms forming space is referred to as "dark matter." In the Theory of Everything, the atom model is singular; therefore, my model does not include elements like protons, neutrons, or quarks found in the conventional model. This makes the theory simpler. However, to avoid conceptual confusion in the future, I have redefined some existing terms within my theory while explaining it, and I will continue to do so. I will not use protons or neutrons, as I could not find a position to equate them within the Theory of Everything. I used the term "neutron" in the context of a neutron star, but this is solely due to the star's established name. Now, I would like to define the electron according to the Theory of Everything.

Electron: In the Theory of Everything, an electron is the moment of a photon's initial formation from an element. When some of the solid atoms forming an element stretch, detach, and cluster away from the element cluster, I call the photon at that moment an "electron." In other words, an electron is another form of a photon, named based on its location, similar to how I defined dark energy as the dark matter regions without elements. Simply put, if there is an element near a soft atom, I will refer to that soft atom or its cluster as an electron.

At the quantum level, a void never forms and never has. To avoid confusion about the concept of a void, let me explain simply: Everything in the quantum field will remain in contact with each other eternally. One reason for this is the negative pressure based on positive force (resistance) from the inside outward, caused by geometric shapes.

Let me illustrate this with a simple experiment you can conduct at home.

Take three sugar cubes and break one of them. The pieces of the broken sugar cube will be smaller than the unbroken ones. (This is why elements are smaller than dark matter particles.) The remaining two sugar cubes are symmetric, while the broken one is divided into asymmetrical pieces. Take a piece of the broken sugar cube and the two whole sugar cubes in your hand. Consider the sugar grains in each cluster as solid atoms: the symmetric sugar cubes represent dark matter, and the small broken piece represents an element. Now, place the asymmetrical sugar piece between the two symmetric sugar cubes and gently press them together without applying too much force. You'll notice that no matter what you do, there will always be a gap on one side around the asymmetrical sugar piece. If you placed a flat piece of paper between the two sugar cubes, there would be no gap; however, natural solid materials never break into flat, paper-like pieces. Since elements form from the fragmentation of dark matter, they have thin and irregular shapes. When elements are placed between dark matter, an absolute void is impossible, so negative force pressure forms in these assumed "gap" areas. Consequently, the solid atoms on the surface of the asymmetrical element expand and become soft. During this expansion, no void forms; the element expands while adhering to the atom on the surface of the dark matter it contacts. You can imagine the surfaces of elements and dark matter like a wet soap bar—slippery on the outside but solid inside. In this state, the element (the asymmetrical sugar piece in the experiment) is in constant vibration because, while its interior remains solid, some of the solid atoms on its surface have stretched like rubber, becoming soft and jelly-like, constantly oscillating. Due to the element's asymmetry, it is impossible to apply equal positive or negative pressure forces from all directions. If more vacuum—i.e., negative pressure—forms around it, some of the atom particles on the surface will fully transform into soft atoms, leaving the space in the form of a photon. This is the formation of an electron. The solid atoms on the surface of dark matter are more compressed and solid compared to those in the element, so the release of soft atoms typically occurs from the solid atom cluster in the element. You can better understand this behavior of the element and electron production by recalling a moment or conducting a simple experiment. For the experiment, you only need a small insect like an ant or a fly—I'll choose a fly because it makes noise. If you've done this before, simply recalling it will suffice. Imagine your hand as dark matter and the fly as an element. Catch the fly with your hand and close your palm. When you squeeze your hand, you apply positive pressure to the fly and the surrounding area; when you loosen your grip, you create negative pressure. Now bring your hand to your ear and listen: When you squeeze your hand, the space narrows, and you won't hear the fly buzzing; when you loosen your grip, the fly finds space to flap its wings, and you'll hear the buzzing. The production of electrons by elements works in a similar mechanism: If dark matter moves in opposite directions, creating negative pressure (the fly buzzes), an electron is produced; if dark matter applies positive pressure toward the element (the fly doesn't buzz), no electron is produced.

The Wave Function of a Photon

If solid atoms are continuously released from an element, they will be consistently observed as electrons. When these releases occur rapidly, a micro-level natural phenomenon known as the "electron's wave function" takes place. As electron emission begins, the soft atoms detaching from the element source cluster together to form photons. Photons leave the element's space in packets and travel at the speed of light through the sequences of dark matter. Dark matter particles, due to the inward-to-outward positive force generated by the photon, move in forced directions to clear a path for the photon. This process results from the detection of electrons escaping the element—rather than the element itself—by devices, leading to the interpretation known as the "wave property of an electron" (similar to Wi-Fi). You can find and examine related visuals online. (Note: As previously explained, directly imaging the element or the photon itself is impossible for the reasons I've outlined.)

Photons are light-mass atom clusters, while elements are heavy-mass atom clusters. Due to their light mass, photons naturally flow at the speed of light. A single photon cluster does not exhibit wave properties, but a wave function emerges during a photon bombardment.

How Does the Photon's Wave Function Occur?

Photons are relatively light compared to elements. The soft atoms forming a photon occupy significantly more volume than solid atoms. For example, the volume occupied by a photon made of 10 soft atoms could, though not definitively, potentially hold millions of solid atoms. Thus, the volume a photon occupies is considerably larger than that of the solid atoms forming an element. You can think of this as the difference in the space occupied by a kilogram of iron versus a kilogram of cotton.

Due to their light and soft/flexible nature, photons are greatly influenced by the pressure of dark matter. This causes photons to add significant kinetic energy to their potential energy. The kinetic energy gained from the surrounding dark matter accelerates the photon in proportion to its volume. The speed and direction of this acceleration are proportional to the positive pressure force and direction of the dark matter the photon is subjected to. For instance, a photon exposed to equal forces from all directions will not move in any specific direction; however, this is a rare phenomenon due to imbalanced pressure zones and asymmetrical elements. If the pressure from the east and west is higher, the photon will rapidly move north or south; if the pressure from the north and south is higher, it will move west or east. I've used directions metaphorically. Considering the hexagonal-cubic structure of dark matter and the 3D environment, countless directional combinations arise during a photon's journey. For example, a photon with the same volume as a dark matter particle travels in a straight line, vibrating (like a laser beam). However, for a photon smaller than a dark matter particle, the distance of its zigzag path increases. During a photon bombardment, photons exit their source consecutively, like the rapid fire of a machine gun. Each photon particle resists the surrounding dark matter from the inside outward during its journey, while the dark matter exerts positive pressure on the photon from the outside inward. This interaction causes the photon bombardment to spread like a wave. Additionally, because photons are volumetrically light, they erode the solid atoms on the surface of the dark matter they pass through during their journey. (Atoms don't erode; by erosion, I mean the softening of the tips of the atoms on the dark matter's surface.) Photons also release a small number of elements from their cluster at the intersection points of dark matter sequences during their journey. The elements detached from the photon cluster create lights—like laser beams or lightning—that allow us to see them as lines even if the photons don't directly reach our eyes. Photons continue their journey unless they collide with an element cluster or the lattice wall of a magnetic field.

The potential and kinetic energy of photons creates a proportional reaction in elements. For example, if you throw a bullet at a thin tree with your hand, its low velocity prevents it from piercing the tree; however, if shot from a gun, it will pierce through. Photons reach element clusters in the form of a bombardment. A single laser beam photon hitting the same spot may not have much effect, but a consecutive photon bombardment produces a noticeable impact. This consecutive photon bombardment is similar to water flowing from a fountain hitting concrete.

The Double-Slit Experiment

When people misdefine natural elements, they may also misinterpret natural phenomena as a result. For instance, in the double-slit experiment, electrons were incorrectly defined according to the Theory of Everything, leading to a dead end. This dead end forced scientists to resort to the so-called "uncertainty principle" as a refuge. However, if the electron had been correctly defined, there would have been no need for the uncertainty principle. The proposition in current science states: "If light were entirely composed of ordinary particles, and these particles were passed through a slit and projected onto a screen, we would observe a pattern dependent on the size and shape of the slit." Here, light is assumed to be ordinary. Yet, in the Theory of Everything, light consists of photons, which are clusters of flexible and soft atom particles. When a single-slit experiment is conducted, the pattern on the screen is a diffraction pattern: a narrow central band in the middle with darker bands aligned parallel to it. This appearance aligns with the outcomes derived from the inductive reasoning of the Theory of Everything.

Similarly, if light were solely composed of particles and passed through two parallel slits, the pattern on the screen would simply be the sum of the patterns created by two single slits. In reality, however, the pattern becomes wider and more detailed, featuring a series of bright and dark bands.

In the large diagram below, you can see the double-slit experiment. Whether there is a single slit at the light source or not, the light spreads in the same way unless it is a laser. (Laser photons, being roughly the same volume as dark matter particles, travel in a straight, vibrating path over a certain distance.)

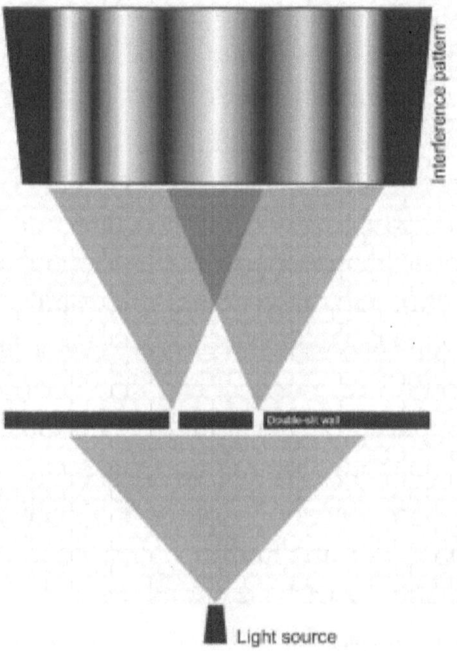

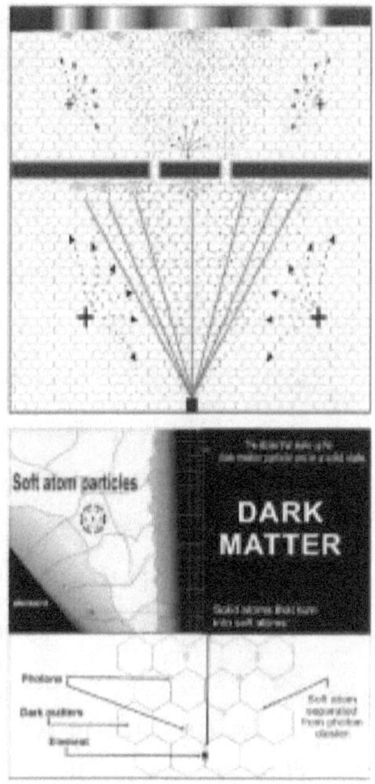

In the Theory of Everything, the electron I have redefined is a cluster of soft atoms that transforms into a photon the moment it detaches from the space of its source element. Photons flow or slide through the spaces between symmetric dark matter particles. During this flow, they deform the tips of the solid atoms on the surface of dark matter due to friction. In the visual at the bottom right, each solid atom that has been subjected to a vacuum has a solid base in its slot while its tips have become soft. Solid atoms completely vacuumed from the element transform into soft atoms. When a photon encounters an element during its journey through dark matter, a violent collision occurs at the quantum level. This collision causes damage to the element; the photon disrupts the structure of the element it strikes and loses elements from its own cluster. (A photon colliding with a cluster of elements can produce numerous different outcomes; here, we are discussing a single element.) The photon's journey through dark matter also causes it to exhibit a wave function, as shown in the diagram at the bottom right.

In the diagram at the top right corner of the image above, the ground appears to be composed of light-colored hexagons, which represent dark matter particles.

Despite its light mass and weak resistance, a photon occupies a considerable volume. Therefore, if you subject a region to photon bombardment, the total photon particles in that region will compress the surrounding dark matter particles (similar to the Karmatan experiment, this compression creates a magnetic field akin to electromagnetism). With its potential and kinetic energy (the positive force pressure applied from the inside outward and the force resulting from acceleration), the photon exerts pressure on dark matter particles. This interaction also occurs with a single photon. For example, when a photon is first formed and passes to the right of an oncoming dark matter particle, it pushes the dark matter to the left. After the photon passes, the dark matter particle, displaced to the left, returns to its original position like a spring, using the kinetic energy gained over the vibration distance. The strengthened potential energy from this kinetic energy allows the next photon to pass on the left rather than the right. This phenomenon has been interpreted in current science as the "entanglement principle" and is one of the factors causing the photon to travel in a zigzag pattern. Let's not forget that these events happen very quickly. Due to the rapid speed of photon bombardment, photons in particle form are also observed and interpreted as waves.

As I mentioned regarding the formation of space, there is no such thing as gravity; everything is related to the positive pressure force of dark matter. Photons, being light relative to their volume, are not easily affected by gravity, celestial thrust, or similar forces. For photons, the primary factor is the pressure in the immediate vicinity of each photon particle. Therefore, the orientation of the double-slit experiment setup—whether toward the sky, ground, right, or left—does not matter to the photon packet. A behavior similar to water from a fountain hitting concrete occurs in all directions during a photon's collision. In the double-slit experiment, most photons pass through the slits as particles, like the flow of a waterfall. Afterward, some of the photons passing through the two slits converge in the middle, creating a wide brightness at the center of the interference pattern. As shown in the diagram at the top right corner of the image above, the white areas—where photons fall densely—are akin to the spots where a waterfall pours. Since many photons fall in these areas, dark matter particles are compressed, causing the spaces between nearby dark matter particles to compress as well. This is one of the reasons for the formation of the interference pattern.

The Wave Function of Water

The wave function of liquids differs from that of light. This is because dark matter particles within water are not arranged in sequences; instead, each dark matter particle in water is surrounded by asymmetrical elements. In the micro-mechanics of water's wave function, the primary factor is, once again, positive force pressure. As you may recall from the liquid state of matter, a liquid has one of the simplest molecular arrangements. The molecular structure of water is quite straightforward. If the small elements around each dark matter particle in a given area are arranged evenly, that cluster (matter) transforms into a liquid state. In liquid states, dark matter particles that have detached from their sequence lose their symmetry and take on a spherical shape. All known and unknown fluid liquids follow this structure. The fluidity of a liquid is proportional to the asymmetrical sizes of the elements surrounding the dark matter particles. If all the elements are small, the liquid becomes superfluid;

if it consists of elements of varying sizes, its fluidity decreases depending on the ratio of large to small elements. Whether it's mercury, liquid oil, or blood, they all share a similar molecular arrangement. However, the mechanics of liquids like lava, molten iron due to heat, or soil differ, as the fluidity of such element groups involves the density of soft atoms. We will address this topic in the heat section.

The wave function of liquids is triggered either by the introduction of an external element cluster into a region of liquid matter or by an imbalance in the dark matter pressure within the liquid's magnetic field. When this external cluster or pressure imbalance affects the cluster forming the liquid, the elements compress the nearest neighboring elements (via positive pressure), initiating an immediate chain reaction. Simultaneously, as a protrusion, elevation, or bulge—essentially a wave crest—forms on the liquid's surface, invisible indentations emerge in the magnetic field lattice on the outer surface of the liquid. Consequently, the asymmetrical, detached dark matter particles forming the liquid (including the symmetric dark matter particles near the liquid's magnetic field lattice wall) and the elements release their potential energy to the nearest surrounding matter, generating kinetic energy. In the wave function of water, the potential energy of each indentation is transferred to the nearest area. At this point, gravity (the direction of the positive pressure force of dark matter) comes into play, and this process repeats until the kinetic energy is absorbed.

Superposition, Colors, and Spectrum

Superposition

Superposition, in quantum mechanics, refers to a particle's ability to exist in multiple possible states or positions simultaneously. This concept is defined by the particle's wave functions. When no measurement is made, the state of a quantum system is expressed as a combination of all possible states. However, upon measurement, the system collapses into one of these possibilities, transitioning into a definite state.

For example, in Schrödinger's cat thought experiment, the cat is in a superposition of being both dead and alive until an observation determines its definite state. This also explains the ability of qubits—the foundation of quantum computers—to be both 0 and 1 at the same time. This is how a quantum event is interpreted according to current theory.

However, I absolutely disagree with the interpretation provided by current science. Why was this interpretation made, and why does the concept of superposition exist?

We all live in nature. In our daily lives or in the past, while working or on vacation, we've likely picked up a stone and thrown it—perhaps at a tree, a wall, a lake, or the sea. Imagine throwing a stone at a wall 10 meters away. Each time you throw the stone toward the wall, it travels through the air and eventually hits the wall. When the stone strikes the wall, interactions and reactions begin in that area. When thrown into a lake, it creates ripples in the water; similarly, a form of ripple occurs on the wall as well. But here's the point I want to emphasize: when the stone hits the wall, the resulting interactions and reactions are not related to you. The moment you throw the stone, all interactions and reactions—including sound and visual information—become independent of you; when the stone hits the wall, the effects are solely between the stone and the wall. Photons behave in the same way. Once a photon leaves its source, it moves independently of that source, and by the time it reaches us or a device, it is entirely independent of its origin. From another perspective, every electron originating from elements becomes free the moment it forms, taking on the name "photon" and embarking on its journey. Now, imagine each stone you throw as a photon and the wall as a light detector. Picture yourself throwing stones at the speed of light, hitting every part of the wall countless times! Photons move so quickly that human perception, overwhelmed by this speed, sometimes defines a property that has no physical example in nature as "both" and "and also." The ability of qubits—the basis of quantum computers—to be both 0 and 1 simultaneously has been explained through superposition in this manner. However, this notion of being both 0 and 1 at the same time stems from a failure to comprehend the speed of a photon/electron. In the Theory of Everything, qubits are either 0 or 1. Let me explain it briefly: a photon/electron is in the 1 position the moment it strikes an element; the moment we wait for the next photon to strike is 0. When a photon hits an element, its potential energy combines with the kinetic energy it carries, resulting in a strong positive force at the moment of collision that pushes the surrounding dark matter particles. After being pushed, the dark matter particles move away and return faster than light, like a spring, over the vibration distance—this is the 0 moment. When interpreted in the Theory of Everything, superposition is not "both, and also" but rather "either, or."

Spectrum and Colors

Information is metaphysical. The reflections of physical events and elements in nature create instantaneous, reactive physical visual images. A photon is the medium that transmits all existing information. When the necessary conditions are met, photons and elements engage in algorithmic physical interactions, enabling the storage of metaphysical information. Whether it's a computer, a notebook, or our brain, it doesn't matter—they are all physical interactions between elements and photons. Photons are a medium that carries visual information. In other words, everything we see with the naked eye or through devices is the result of reactions in our brain triggered by photon clusters colliding with elements.

Light photons travel through the spaces between dark matter particles. If the elements between dark matter particles are sparse (for example, nearly absent in the vacuum of space or as sparse as in the gaseous state of matter), the light photon continues its journey without losing much mass (its cluster of soft atoms). However, if it encounters an element during its journey, the light photon cluster disperses depending on the size of the element. When the photon cluster disperses, it forms multiple new photon clusters with fewer soft atoms, and each cluster can be perceived as different colors by devices or living beings. A rainbow is an example of this phenomenon. This is referred to as the spectrum of light; this is how a spectrum forms.

Colors do not exist in nature. More precisely, the entirety of space is filled with clusters of atoms, with over 99% consisting of dark matter/energy, photons, and elements. Colors are created by our brain. The fact that our brain generates colors indicates that it is undergoing an interaction. This means that before our brain processes colors, photons enter the magnetic field of our eyes and collide with the elements that form the organic detector of the eye.

The photons detected by the elements in our eyes, which we can think of as sensors, are processed in our brain independently of their source—similar to how a wall is independent of the person who threw a stone at it. In conclusion, all colors are metaphysical, and every living being or device can produce colors based on the photons they perceive. Countless beings, such as colorblind humans, insects, and fish, perceive colors differently or struggle to distinguish them.

Reflections and Mirrors

We have learned that every element cluster has a magnetic field. From a virus to a galaxy, everything that forms the universe has a magnetic field, and anything with a magnetic field also has a magnetic field lattice. This lattice has a solid wall, the strength of which depends on the density of the dark matter particles. In this context, let's start by explaining how mirrors reflect light. Photons, being volumetrically light in mass and moving due to dark matter pressure, can only continue their journey without deviation if there is a clear path. However, when they encounter an object composed of an element cluster, the first obstacle they face is the object's magnetic field and the wall of its magnetic field lattice. Indeed, as you may recall from galaxy formation, the magnetic lattice can sometimes trap and reflect a significant portion of photons escaping from the inside outward, while also blocking external photons depending on the wall's strength. The magnetic lattice wall of a mirror is flat and solid, primarily due to the molecular arrangement of the elements forming the mirror. The structure and arrangement of the elements determine the shape of the magnetic field lattice wall. These walls can be flat or have protrusions and indentations. Walls with protrusions and indentations are used in stealth technology to avoid radar detection, while a mirror's wall is flat. As a result, a photon cluster hitting a mirror is largely reflected back along its axis depending on the angle of incidence. This is why, when we look into a mirror, we can see ourselves and other objects at an angle. The sole reason for this is that the photon cluster bouncing off the magnetic wall reflects back without losing many elements. Sometimes, the magnetic field lattice wall on the surface of water can also take on a solid and flat shape; this is why water can reflect like a mirror.

We, or devices, can only see an element cluster through photons. Photons carry visual information from the object we see in two ways: either they originate from another source, hit the object, and reach us, or they form directly at the source and come to the observer. Regardless of how the photon reaches the observer, the magnetic field factor plays a significant role.

In mirrors or reflective surfaces, the molecular arrangement of elements is dense and solidified. Therefore, objects like mirrors that reflect light extensively are not sources of the photons that form visual information. For example, if you look at a mirror on the wall in a pitch-black room, you won't see the mirror or yourself because a mirror is not a photon source. However, if you light a candle in that dark room, you'll see the candle's light; the flame from the burning part of the candle is a light source.

Before moving on to colors, by discussing mirrors, we have learned about the effects of magnetic fields and the difference between a light source and something that is not a light source.

The Formation of Colors

Colors are a vast and comprehensive topic. In the Theory of Everything, it is impossible to fit everything into a single book. Therefore, it is more efficient to explain some topics briefly and move on to the next. The purpose of this book is to describe natural elements and their operational mechanics according to the Theory of Everything. In this context, the formation of colors can be summarized as follows: Colors depend on the element structure of the object, the arrangement of elements, the surfaces of dark matter deformed by the molecular arrangement of elements, the shape of the object's magnetic field, and the strength of its magnetic lattice wall. All these factors cause changes in the color spectrum of the reflected photons. Colors can also form directly at a light source. However, the final form of colors is determined in our brain after passing through our eyes.

The visible light spectrum consists of different light frequencies, and these frequencies determine the color we see. The wavelength of light determines its frequency. The wavelength range is fundamentally proportional to the number of soft atoms in the photon cluster. As the number of soft atoms in the photon cluster increases, the wavelength range shortens because a photon cluster with more soft atoms is more affected by the pressure force of dark matter, accelerating and shortening its wavelength. Conversely, if the photon cluster is volumetrically small, the dark matter's pressure provides less acceleration, causing the photon to move more slowly and resulting in a longer wavelength. Thus, the speed of light is not constant; it depends on the volume of the photon, which in turn determines the colors. To make this clearer, let's classify colors based on the number of soft atoms in the photon cluster. Suppose a photon forms on the Sun and travels toward Earth, consisting of a cluster of 100 million soft atoms. As the photon travels through space between dark matter particles, it loses elements over the distance (let's say it loses 3 million elements) and enters Earth's magnetic field. Upon entering the magnetic field, it loses another 10 million soft atoms. It continues traveling and then collides with Earth's magnetic lattice wall. At this moment of collision, the photon cluster might not lose any elements because it has moved from an area of high dark matter pressure to one of low dark matter pressure. When it left the Sun, the photon cluster had 100 million soft atoms; now, 87 million remain. If it loses another 30 million elements while interacting with the elements forming the air, 57 million elements remain. Finally, the photon reaching the Earth's surface collides with a flower, fragmenting further, and some photons travel toward our eyes. Let's say that after hitting the flower, the photon that reaches us, originally with 100 million elements on the Sun, now has 7 million elements left. This cluster of 7 million soft atoms loses more elements in the magnetic field on the surface of our eye and passes through the elements forming our eye, triggering an element that initiates an algorithmic process. At the end of this long journey, we perceive the color reflected from the flower as a pixel. From this journey, it's clear that the ultimate place where all colors are determined begins in the magnetic field of our eye. Just as there can be eyes with a wide magnetic field and a weak magnetic lattice wall in large eyes, or a strong magnetic lattice wall in small eyes—or vice versa, with countless combinations—there can also be eyes formed by various molecular arrangement combinations of countless elements. Additionally, the brain of each living being may have different algorithmic processing combinations that determine the image.

In the Theory of Everything, examining life at the quantum level can help us better understand some of biology's most fundamental processes. I will briefly explain how processes like photosynthesis, viruses, consciousness, and even information transmission in our nervous system operate through quantum mechanical effects in this theory. This topic, like others in the book, is vast and comprehensive. Still, I believe this perspective on life at the quantum level will bring a fresh outlook to biology's core processes.

According to the Theory of Everything, everything in nature is made of indivisible atoms. From the structure of black holes to the cores of planets and stars, indivisible atoms form the foundation of all things. Despite this, the emergence of countless different objects from these atoms never ceases to amaze. Yet, the atom—the fundamental building block of everything—is remarkably simple. Indivisible atoms exist in two extreme states: solid and soft. Solid atoms form symmetric dark matter and the fragmented, asymmetric elements derived from dark matter. When solid atoms expand and transition into a soft state, clustering together, they form photons. Thus, we can classify the quantum realm of nature into three main categories:

- Symmetric dark matter/energy,
- Asymmetric elements,
- Asymmetric photons (asymmetric due to the countless combinations of soft atoms in clusters).

The interactions between these three categories explain everything in nature, including life.

To understand life, we must first explore viruses. In the Theory of Everything, viruses are clusters of elements, each made of solid atoms. These elements consist of small, asymmetric shapes and exist between dark matter, bonded together. For example, a triangular prism-shaped oxygen element might be bonded to a cube-shaped hydrogen, and both hydrogen and oxygen may be chained to a larger carbon element. These element chains can sometimes extend in spiral shapes (like DNA sequences) between dark matter. The atoms on the surface of these elements are usually in a solid state. Such solid chains of light elements—viruses—are found everywhere in the universe. The number of viruses in planets, stars, meteorites, comets, and asteroids depends on the level of photon bombardment and the vibration frequency of elements in that environment. For instance, in stars, where the vibration frequency of elements and photons is high, viruses are nearly nonexistent, and their static duration is short. On Mars, however, the number of viruses is high, and their static duration is long. On Earth, the static duration of viruses varies with environmental activity, and there are fewer viruses than on Mars. Why? Because most viruses on Earth have transformed into components of life, such as bacteria and cells.

If we compare viruses to something in our own scale, they resemble a hardened cookie. A cookie—like many solid foods—is essentially a cluster of viruses and is inanimate. But if we dip the cookie in tea, it becomes more "alive" compared to its previous state (by "alive," I mean transitioning from a static to a dynamic state). When viruses—clusters of solid elements—are exposed to photon bombardment, they "dissolve," much like a cookie softening in tea. The atoms on the surface of these solid elements expand through a vacuum, creating a dynamic, "alive" environment. This can lead to an explosion of life on a planet. All it takes is a stable photon frequency over a prolonged period. For example, if Mars were exposed to a stable photon frequency for a long time, a visible explosion of life would occur within a few months. Within decades, Mars could transform into a planet with seas, lakes, rivers, trees, plants, and animals. However, this isn't possible under current conditions because the photon frequency Mars receives is much lower than Earth's. To recall my earlier explanation about photons in the context of colors: if a photon leaves the Sun with 100 million elements and arrives at Earth with 87 million, it reaches Mars' magnetic wall with even fewer elements. As I've explained, many factors affect a photon's frequency.

Let's continue exploring viruses. Viruses are entirely composed of solid elements, which are asymmetric. There are no soft or slightly expanded solid atoms between the molecular bonds or on the surfaces of the elements that form viruses. Viruses are akin to a dried, hardened piece of soil. Humans live closely with viruses—there's no escaping them in daily life. For instance, the particles we inhale while smelling are viruses. All the dry foods we eat are made of viruses. Some of these virus-based solid foods are chemically processed and separated to make meals, while others we eat directly, letting our bodies break them down.

Before explaining the transition from viruses to life, I'd like to reframe the statement "Cells multiply by dividing." Yes, cells multiply through division: one cell divides into two, and countless cells dividing countless times result in countless cells. When these countless cells come together, they form a single structure. I liken this to the sea: each water droplet is a cell of the sea, and countless droplets form a sea-sized body of liquid. Humans are the same. If we take a piece of a human, we encounter cells, and countless cell clusters form a human. With this understanding, I'd like to propose the following: "In an organism that has formed and become human, cells do not multiply by dividing." This is because the cells are already united, forming a single structure—the human. In a single structure, cells cannot multiply by dividing, much like how a fixed number of water droplets forming a sea cannot multiply by dividing.

Let's explore the emergence of life, starting with viruses. Now that we understand viruses, we can predict that when solid viruses are exposed to appropriate photon bombardment, they will break apart, and the atoms on the surface of their elements will expand. This is similar to the process I previously described for the transition of matter from solid to liquid. During bombardment, photons collide with the solid elements between dark matter, disrupting their structure by creating negative pressure on the element's surface. As a result, the element cluster, which occupies little space in its viral state, expands and takes up more space. This expansion detaches the dark matter from its array, wrapping around it like a ball bearing. Consequently, the elements forming the virus transition from a solid to a liquid state, creating the first cell.

Let's consider this first cell as the first human cell. I'll place the emergence of humans in the phase of the first life explosion after the mass extinction that ended the dinosaur era. Although I refer to it as the "first cell," countless different cells formed simultaneously in countless places on Earth. This is because the extinction of the dinosaur era led to the death of countless dinosaurs and other creatures, reducing or entirely eliminating the predator-prey populations on a macro scale. Let's return to our first cell.

The cell, which transformed from a solid virus by breaking apart, is in a liquid state. It assimilates suitable elements from the outside while rejecting unsuitable ones. This process occurs thanks to a magnetic field. The magnetic lattice surrounding the cell acts like a filter, preventing large elements from easily entering. The internal pressure pushing outward in the cell balances the external pressure pushing inward. As a result, the cell begins to grow—note that it grows as a single piece, not by dividing. (I previously explained that in liquids, the magnetic field can be weak or strong depending on environmental factors.) Growing as a single piece, the cell filters and assimilates external materials while withstanding the external pressure from dark matter. If it cannot withstand this pressure, it will split in two. As materials continuously enter the cell from outside, the internal environment remains dynamic. The elements inside the cell cluster based on their shapes, forming smaller clusters within the cell. At the same time, a skeletal structure begins to form. (The primary factor here is geometric shapes and sizes: symmetric and asymmetric elements push the minority outward while gathering the majority at the center—or vice versa.) The skeletal structure forms in the central parts of the cell, becoming more solid than other clusters due to pressure-induced synthesis. As the initially liquid cell grows and incorporates solid elements, it loses its fluidity and forms a skin that encloses its outer surface. Now, the cell has reached a visible size. Since most of its surface is covered by skin, element intake and the expulsion of undigested elements (waste) occur in a single direction. The shape of this organism resembles a primate—specifically, a human.

Since the life explosion occurred simultaneously in countless locations on Earth, millions, perhaps billions, of the same species formed from this virus-to-cell transformation. In other words, humanity does not have a single common ancestor. In the Theory of Everything, there are no processes like the transition from fish to human. When humans first emerged, they were tiny, almost microscopic. As a species, they survived and grew to their current sizes over time.

This is a hypothesis. I cannot prove that life emerged this way, but in the future, experts well-versed in the Theory of Everything may unravel the mysteries of all living beings by adhering to its laws. Perhaps someone could test my hypothesis using a computer simulation based on the laws of this theory. According to the Theory of Everything, every species currently living on Earth is not much different from how it was when it first emerged. During the life explosion, every creature grew from a tiny size to its adult form. In other words, humans, dogs, elephants, giraffes, whales—all living beings—reached their current sizes years after emerging from a finger-sized form.

WHAT IS ELECTRICITY?

Electricity: The Pulse of Life

Electricity is an indispensable part of modern life. From homes to workplaces, transportation to industrial facilities, electrical energy is used everywhere. This form of energy simplifies daily life and forms the foundation of many technological advancements. The discovery and use of electricity marked a major turning point for humanity. Beyond meeting basic needs like lighting and heating, electricity powers electronic devices, enables communication networks, and enhances the efficiency of production processes.

The fundamental building block of electricity is the soft atoms that form photons—not the solid state of atoms, but their soft state. In the Theory of Everything, we know that soft atoms cluster to form photons. In electricity, however, soft atoms are temporarily arranged in a fixed pattern around an element. We previously defined an electron as follows: In the Theory of Everything, an electron is the initial moment of a photon's formation from an element. When some solid atoms forming an element expand, break away from the element cluster, and shift outward as a cluster, we call this a photon. But the soft atoms that cannot escape and remain around the element—or those that slide between elements by pushing other electrons—become the potential electrical form of the electron. Notice I said "potential": As long as soft atoms remain around the element, the electricity phenomenon does not occur. In this sense, most objects in nature hold potential electrical energy. We call such objects "conductors," and others "insulators." The distinguishing feature between conductors and insulators lies in whether the atoms around the element's surface are in a soft or solid state. If the atoms (depending on their density) are in a soft, expanded state, the object is a conductor; if they are in a solid state, it is an insulator. This varies based on the element's structure and molecular bonds, which is why not all solid or liquid objects are conductors. For example, plastic, glass, wood, Teflon, porcelain, paper, ceramics, mica, rubber, Bakelite, pure water, sugary water, and gases are insulators. Metals (such as copper, iron, gold, silver, aluminum, lead, platinum, zinc), graphite, saltwater, acidic water, and the human body are examples of conductors. Conductivity or insulation depends on the element's structure and molecular arrangement. In some conductors, electricity flows directly through the object, while in others, it flows around the outside. In the future, researchers may use the Theory of Everything as a foundation to write theses on the element structures and molecular arrangements of all objects.

As we continue to explore electricity, let's understand why gases are insulators and the difference between photons and electricity, both of which are made of soft atoms. If light is made of the same atoms as electricity, why doesn't light shock us like electricity does? Let me explain with an analogy: Imagine the sky as it truly is—covered with dark matter. Now, picture the raindrops falling from the sky as photons. When it rains, there's a gap between the raindrops. For example, on a rainy day, hold your index finger steady. After one raindrop lands on your finger, you'll need to wait a few seconds for the next one to fall. The reason light photons don't shock us like electricity—and why gases are insulators—lies in the time gap between those two raindrops hitting your finger. When a raindrop hits your finger, it pierces the invisible magnetic lattice wall on your skin and interacts with the elements in your skin, an interaction we feel in our brain. Since the magnetic lattice wall is made of dark matter, when the raindrop hits, the pierced wall reacts with the potential energy of the dark matter and closes again. To make this clearer, let's use another analogy within the analogy: Think of a motion-sensor door at a shopping mall—the kind that opens when you approach and closes after you pass through. Imagine the door as the magnetic lattice wall and the people passing through as photons. When people pass one by one, there's no issue, but if dozens of people try to pass at once, problems arise. Electricity works similarly. When we stand under the Sun, countless photons hit us. Each time they hit, the magnetic lattice wall opens, but it closes faster than the speed of light. Due to the frequency range or the time between waves, an electric shock doesn't occur. However, when we touch a live electrical cable with our bare hand, all the soft atoms between the elements forming the cable flow through the magnetic lattice wall—already opened by our touch—into our body. During this flow, the potential and kinetic energy of each soft atom compresses everything in our body with positive force, causing reactions. Simply put: we get shocked.

One of the key factors in insulators and conductors is the frequency range. If soft atoms reach their target in clusters with gaps or wave-like properties, they don't cause a shock. But if they flow in a continuous stream, a shock occurs.

While on the topic of electricity, let me correct a misconception using the Theory of Everything: Lightning, thunderbolts, and the mini lightning from a Tesla coil are not electricity. Therefore, don't bother researching ways to generate or store electricity using them. Lightning and thunderbolts are not the electrical form of soft atoms but their photon form. In other words, lightning and thunderbolts are photon clusters made of soft atoms. When lightning strikes a person, the harm comes from the high kinetic energy due to the lightning photon's large size and speed.

The raw material of both photons and electricity—soft atoms—has always formed in the same way. Soft atoms are created solely through the negative pressure resulting from a chain of positive pressure-based interactions, as solid atoms stretch and expand into soft atoms. This is a physical law in the Theory of Everything. Whether it's lighting a match, the Big Bang that formed the universe, or a supernova explosion, the formation of soft atoms remains the same. This physical law also applies to lightning and thunderbolts. When you strike two pebbles against each other in opposite directions, you use positive force, resulting in negative pressure at the collision point. The spark that emerges is a tiny thunderbolt—a much smaller photon compared to a lightning photon. Lightning and thunderbolts form in the same way. In the sky, elements form clouds. Clouds become macro objects made of micro-level dark matter and elements. Every macro object has a magnetic field. In the sky—whether it's clear or cloudy—there are invisible macro dark matter cells with magnetic fields. During cloudy weather, due to the high density of elements, these invisible macro cells can generate stronger magnetic fields and positive force. When these macro cells collide at a small point—like the pebbles in our example—negative pressure emerges. At the center of this negative pressure, the solid atoms forming the dark matter or element are vacuumed, creating a large photon cluster of soft atoms. This cluster is the lightning or thunderbolt photon. The lightning cluster, trapped between dark matter, moves toward the direction of lower pressure due to the strong pressure of the dark matter. The route of this journey is determined by the positive pressure force. The lightning photon, with its momentum, collides with something and shatters. A thunderbolt photon, however, disintegrates in the sky and disappears. Let's assume a thunderbolt or lightning forms with a 5 cm diameter. As the photon cluster moves between dark matter arrays, it loses elements at each intersection point. (These lost elements are the light photons that reach our eyes from the lightning or thunderbolt.) Since the distance between the lightning and the ground is short, it doesn't lose many elements before colliding and shattering. A thunderbolt, however, travels through the sky, releasing light photons in all directions from its initial 5 cm diameter until all its elements are depleted, and then it disappears. I hope this explains why electricity cannot be generated from thunderbolt and lightning photons. Thunderbolt and lightning photons either shatter upon impact or turn into light shortly after forming. Due to their large size at formation, their momentum is high, making them impossible to capture.

While discussing thunderbolts and lightning, let me also explain the quantum mechanics of solar panels. This will help you understand electricity generation as well.

HOW IS ELECTRICITY GENERATED AND STORED?

In the Theory of Everything, we call the soft atoms that either expand from a solid state to a soft state through negative pressure from an element or position themselves near an element from elsewhere "electrons." When electrons break away from the element and travel through dark matter, they transform into photons. The flow of electrons between elements, like a river, is what we call the electrical form of electrons. In the Theory of Everything, all concepts are simple, testable through experiments, and relatable to everyday life. Keep the analogy I'm about to share in mind to better understand electricity. For this analogy, imagine a setting like a prison corridor with cells on both sides or a university hallway with rooms on either side. The corridor in your chosen setting represents an electrical cable made of elements. The walls of the rooms are dark matter, and the people inside the rooms are the electron form of soft atoms. The outside of the corridor and rooms is the insulating sky. This analogy can also apply to a stationary train on a railway with wagons and a corridor on both sides—however you'd like to picture it.

Let's continue the analogy with a train. The train has a long, narrow corridor where only one person can barely walk. On both sides of the corridor are 50 wagons, each designed for 2 people but capable of holding more if crowded. At the back of the train, there's an entrance door from the outside, and at the front, there's an exit door. Since the train has a capacity of 100 people, if 1 person enters the back wagon from outside, the person in that wagon moves to the next wagon, and this continues until someone exits from the front wagon. The flow of electricity works in a similar way. In an electrical cable, if a soft atom enters from one end, it pushes all the soft atoms ahead of it, and a soft atom is expelled from the other end of the cable. However, a soft atom cannot enter or exit the cable without an additional mechanism due to the cable's magnetic lattice wall. This is how electrical grids are designed and operated. From nuclear power plants to dams, all known grids provide electricity based on this principle. For example, electricity is generated at a dam 50 km away. At the dam, soft atoms are pushed through the cable that forms the grid. (This topic will be detailed in the hypotheses on direct and alternating current.) In your home, soft atoms exit from the screen of a plugged-in, working television, enter your eyes, and create color pixels in your brain. Everything else is just technical detail.

Now that we understand electrical current, let me explain how solar panels generate electricity using the elements of our analogy. This time, imagine the train not as a cable but as a flat panel that converts sunlight into electricity! The materials that form light photons and electricity are fundamentally the same: soft atoms. I previously explained that electricity doesn't form in wave-like photons due to the magnetic lattice wall. In our analogy, the train is now flat or consists of thousands of trains lined up side by side. Its surface is like a solar panel, but the hypothetical elements representing electrons and elements remain the same. Let's say a photon cluster from the Sun consists of 3 soft atoms (3 people). When the photon hits the panel with the momentum generated by its velocity, the panel's magnetic lattice wall opens and closes instantly. During this moment, one of the 3 soft atoms stays inside the lattice, while the other two reflect in different directions. Thus, an external soft atom joins the panel cluster. If we imagine the soft atom as a person, this external person entering the train causes someone from the exit wagon to leave. The person who exits—or the soft atom—immediately functions as electrical energy through a cable. In other words, the person exiting the train pushes the last person into a storage unit, which we'll call a battery. This is how solar panels generate electricity: They capture soft atoms from countless wave-like photons and trap the soft atom pushed out from the exit. If the exit door for the incoming soft atoms is closed, the rooms will be full, and new soft atoms cannot join the cluster.

Storing electricity in devices like batteries is easier to explain. Imagine the train as a battery with the same elements. If you close the train's exit door and fill the rooms with more than 2 people instead of just 2, you've stored electricity. As always, the structure and molecular arrangement of elements play a significant role in electricity storage.

INSTINCTS

What Are Instincts?

Instincts are natural impulses that drive living beings to perform beneficial or necessary actions without the involvement of reasoning, thought, or consciousness.

I hypothesize that the foundation of instincts begins when a living being takes on its cellular form. I previously explained how life begins in the Theory of Everything. The world of microscopic organisms is vastly different from that of macroscopic ones. All living beings, including elephants and giraffes, were once microscopic in size. For example, humans start as microscopic entities in the womb. During the life explosion, when no living being had a mother, countless organisms emerged in countless places as viruses transitioned from a solid to a liquid state.

Instincts start forming in the cellular stage. A cell transitioning from a solid virus form to a liquid form under the influence of photon frequency reacts to external elements and photon frequencies entering its cluster. Let me illustrate this with an experiment: If you place a smaller water droplet next to a larger one on a flat surface and repeat this in a chain, the larger droplet moves in that direction. In the wild, similar events after a cell forms lay the foundation for instincts. Depending on the elements forming the cell's structure, the cell assimilates symmetric elements and grows in that direction. However, external photons with short wavelengths or large volumes can disrupt or break the cell's structure or push it in the opposite direction. Even today, most microscopic organisms on Earth—typically insects—move in this way. Tiny creatures like ants, flies, and bees move and communicate based on the frequency and structure of photons and viruses. Even conscious macroscopic beings like humans sometimes cannot resist the effects of photon and virus frequencies. Colors formed by photon frequencies trigger instinctual reactions in our multicellular structures, making something appear beautiful, ugly, delicious, poisonous, cute, or scary to us. When lightweight virus clusters of solid elements float in the air, the scent we detect through our nose causes reactions in our multicellular structures. This instinctually determines whether the organism moves toward or away from that scent.

What Is Scent? How Does It Form?

All scents are element clusters in the form of viruses. Structures made of asymmetric lightweight elements give rise to life when the right environmental conditions are met. Living beings feed on viruses throughout their lives. Depending on the organism's genetic makeup, viruses can be beneficial or harmful. Viruses entering an organism break apart, and each element in the solid element chains forming the virus finds its place within the organism's dynamic internal structure. Some assimilated elements bond together in a solid state again, forming structures like shells, skeletons, hair, or teeth. However, most elements are expelled from the body due to internal-to-external positive pressure, and these expelled elements create scents. For example, humans constantly shed microscopic skin particles, which are viruses. The shedding of skin occurs rapidly depending on the movement of internal elements and electrons. For instance, during high-effort activities, the movement of elements and electrons in the blood increases. As this movement intensifies, the number of soft atoms (electrons) also rises. Recalling that soft atoms form by expanding from solid atoms, the internal-to-external pressure in the body increases. This pressure surge causes the rapid formation and expulsion of solid element chains (skin) on the body's outer surface. As skin shedding accelerates, the simultaneously expelled soft atom clusters condense the skin into sweat. In conclusion, scent is a form created by the synthesis of lightweight elements in their solid state. The synthesis of solid elements can result in countless molecular arrangements. We call these synthesized lightweight solid element chains viruses, which can be harmful or beneficial depending on the organism's genetic makeup.

How Does Communication Occur?

Every topic I delve into becomes quite broad and comprehensive. The Theory of Everything can explain all mysteries in nature. However, I cannot fit everything into a single book, which is why I often repeat certain points. My hope is that by thoroughly learning the properties of natural elements in the Theory of Everything, you'll be able to solve the mysteries of all natural phenomena in seconds.

Everything in nature is made of atoms. Atoms have two extreme states: a solid, compact, small-volume, indivisible raw state, and a soft, expanded, large-volume, indivisible state. Based on the properties I've defined for atoms, I can make this analogy for elements: "Solid atoms are the inn, soft atoms are the travelers." From sound to Wi-Fi communication, all forms of communication in nature are facilitated by soft atoms. Clusters or entities made of solid atoms serve merely as the path or the stop. Now that you know what enables communication, let me briefly explain the quantum mechanics of one type of communication.

HOW DOES THE INTERNET WORK?

In nature, no tangible entity can be copied. Copying only occurs by replicating the process of an event. All information arises from the reflection of quantum-level physical events into our macroscopic world. Since information is the reflection of an event, it can be copied. In other words, you can copy information, but you cannot copy the atoms at the source where the information originates. Let me explain copying further: Imagine you're at home with a friend, and there are domino tiles on the coffee table. Your friend arranges the dominoes close to each other in a line. You see this and do the same. Then, your friend pushes the first domino, causing them all to fall in sequence. When you do the same, you've copied the event. Your friend "copied" the action by knocking down the dominoes, and you "pasted" it by knocking down your own dominoes. In computers, the "copy-paste" function works the same way. Tangible objects cannot be copied; the only thing that can be copied is information.

I have no expertise in any field, and my past profession, in which I was skilled, is unrelated to this topic. I've never studied any scientific discipline, including mathematics. With only an elementary school education, I'm trying to describe nature in this book. Mathematics is a human invention, and its foundation, even if zero wasn't initially recognized, is built on 1 and 0. So, what are 1 and 0? In the Theory of Everything, 1 represents, from smallest to largest: an indivisible solid atom, an indivisible soft atom, an asymmetric solid atom cluster (element), an asymmetric soft atom cluster (photon), clusters of elements, star systems, a cluster of star systems (galaxy), a cluster of galaxies (universe), and ultimately the space that contains the universe. In the Theory of Everything, "One is everything, and everything is one." If I've explained what 1 means, then what is 0? In the Theory of Everything, the definition of 0 is simple: Zero is a boundary. In natural entities, 1 is a cluster or a single piece, while 0 is the boundary of that cluster or piece. In natural events, 1 is the event itself, and 0 is the time between the end of one event and the start of another. All communication in nature occurs through entities and the events they produce. For example, in communication, 1 is a photon or an electron (the photon form near an element), and 0 is the start and end of the event involving the photon or electron—or the time between two events in repeated occurrences. In this context, I could even declare 1 and 0 as a physical law. The scientific community can designate it as a physical law in the future if they see fit.

Now that we understand 1 and 0, let's explore algorithms, which play a crucial role in how the internet works. An algorithm is a set of step-by-step instructions designed to solve a specific problem or complete a particular task. These instructions are organized in a logical sequence, with each step clearly defined. Algorithms are used across a wide range of fields, from mathematics to computer science, and even in simple daily tasks. For example, a cooking recipe is an algorithm: It specifies step by step which ingredients to use, how to mix them, and how long to cook. In computer programming, algorithms define how a computer performs a specific operation.

Simply put, an algorithm is a process with an end. Algorithmic systems also form the basis of how our brain works. Whether active or not, our brain constantly performs algorithmic processes. By "algorithm" in the brain, I mean the finite effects and reactions at the quantum level. For example, when someone asks, "Hello, what's your name?" the photon-based sound frequency is processed in your ear, triggering algorithmic (finite) effects and reactions between the elements and electrons in your brain. When you respond, the muscles in your brain initiate another algorithmic process through positive force on the elements and electrons. Computers and the internet operate on the same algorithmic principles. When you turn on a lamp at home, the bulb lighting up is algorithmic. Swiping a photo on your phone is also an algorithmic process.

The internet is a communication tool that works by sending and receiving soft atoms in the form of electrons or photons, encoded as 1s and 0s in waves. At the quantum level, this is how the internet functions. All communication tools, including the internet, operate on the same principle, and natural communication is no different. If photons and electrons don't travel in a wave function, communication doesn't occur—this includes visual and sensory communication. We previously learned that electricity is not a wave but a direct flow of soft atoms. Yet, internet cables and computers operate with electricity. You might ask, "How does this work?" The answer: Semiconductors. Semiconductors function as you've learned about the wave function of electrons. (We can also create algorithmic processes by determining voltage levels.) When an electron passes to the right of dark matter, the dark matter compresses to the left, vibrating with kinetic energy. Semiconductor element arrays work the same way, converting the flowing electricity—like water—into a wave form. It's like turning a faucet on and off: If you open the faucet, the water (electricity) flows steadily; if you turn it on and off, it flows in waves.

In conclusion, at the quantum level, the internet operates through a coding/encoding system of 1s and 0s. In Wi-Fi transmission, the wave system is facilitated by photons, while in cable transmission, electrons are converted into a wave function through semiconductor valves.

WHAT ARE HEAT AND TIME?

In current science, heat is defined as a form of energy in thermodynamics, transferred from one substance to another due to a temperature difference. At the microscopic level, heat is related to the motion and vibration of atoms and molecules. It occurs as an energy flow from a warmer substance to a cooler one, continuing until thermal equilibrium is reached. As a form of energy transfer, heat is measured as the amount of energy transferred per unit of time and is expressed in joules (J). Time, on the other hand, is a concept used to measure and organize the sequence, duration, and changes of events. In physics, time is typically considered one of the four fundamental dimensions and is addressed within the framework of space-time. Time plays a significant role in both daily life and scientific research.

I don't have much objection to the definition of heat, but I disagree with the notion of time being a dimension. I'd like to examine these concepts through the lens of the Theory of Everything and present my ideas.

Heat: Heat is a form of energy that emerges and is perceived at the macro level. In the Theory of Everything, at the quantum level, there exists only a single force and energy: the positive force, born of resistance, which renders the destruction of potential energy impossible and acts from the inside outward. All forms of energy, including thermal/heat energy, are physical perceptions resulting from the interactions of atoms. Everything perceived by humans stems from changes occurring at the quantum level. At this quantum level, the only thing that happens is a change in the position and/or volume of atoms or atom clusters. The expansion of a solid, compact, indivisible atom initiates an irreversible entropy. Given the countless atoms in space, as you may recall from the Big Bang, the transition of just a few atoms from a solid to a soft state—or their expansion—triggered entropy and led to the formation of the universe. The ongoing transition of countless individual atoms from solid to soft in numerous places continues to sustain entropy. As a result of this paragraph, all energies we perceive, including heat, as well as time, are part of an ongoing entropy that will never cease for eternity. (The fundamental reason for entropy is that soft atoms, formed by the expansion of solid atoms, can never be compressed back to their smaller form. Compressing an expanded soft atom back into a solid form, whether naturally or artificially, is impossible because a single soft atom, when compressed, will slip away through dark matter or between elements to another location.)

In the Theory of Everything, all events in nature are quantum-based, so there is no difference between macro and micro physical events. In other words, all laws in macro physics apply to micro physics, and all laws in micro physics apply to macro physics. For example, all classical physics laws—from thermodynamics to electromagnetism and optics—can find a place within the quantum framework of the Theory of Everything.

The physics in the Theory of Everything is very simple. Recall the analogy of a fly in your palm: If you squeeze something, the surrounding space shrinks, reducing its range of motion; if you loosen it, the range of motion expands. When the range of motion expands, vibration increases, leading to expansion. The perceptions of heat and time arise in this way. Distinguishing between metaphysics and physics correctly is highly beneficial. To determine whether a concept is metaphysical or physical, it's enough to check if it involves paradoxes and/or relativity. For instance, you might conclude, "Temperature and colors involve relativity, while time involves both relativity and paradoxes; therefore, they are metaphysical."

At the quantum level of nature, there is no hot or cold. An atom is either in a solid state or a soft state. In regions with soft atoms, temperature emerges at the macro level due to vibration. The most critical condition for temperature is the presence of elements between symmetric dark matter. Let me give two examples: The cores of the Sun and Earth are neither hot nor cold; heat energy does not form in their cores. This is because the internal structures of the Sun and Earth's cores consist solely of soft atom clusters (photons). Black holes are also neither hot nor cold because they are made only of dark energy, another form of dark matter. Since everything, including humans, is made of elements, if there are no elements, the perception of temperature—a unit of heat energy measurement—cannot form in devices or living beings.

TIME

In the Theory of Everything, time is a unit of measurement—like weight, length, and temperature—that living beings can perceive. Humans, being more intelligent than other creatures, have divided time into years, hours, and minutes, using it in mathematical calculations. Animals are also aware of time, just as they are of temperature and length, but they use it in a simpler way. Many animals can sense when morning or evening will arrive. The ability to perceive time is unique to creatures with brains. Our brain performs countless algorithmic processes every day, and we perceive time relative to the intensity of these processes. For those who are constantly thinking or working, time passes quickly; for those who don't think much, time moves more slowly. Time is metaphysical and relative to the number of processes in a person's brain. Dreams like time travel or stopping time are not science fiction but mystical fantasies. However, because time is metaphysical, metaphorical time travel is possible: When you sleep at night and wake up in the morning, you've traveled forward in time; when you recall past moments in your imagination, you've traveled backward in time.

HOW DO THE HEART, NERVOUS SYSTEM, AND BRAIN WORK? WHAT IS CONSCIOUSNESS?

Let me remind you again that I have no expertise in any field. I'm trying to explain the Theory of Everything by writing a book. Writing this book took years, giving me ample time to think deeply about every topic. Although many topics in the book are broad and comprehensive, I've summarized them briefly. Going into detail wouldn't fit in a single book—it would take hundreds of volumes, like an encyclopedia. The heart, brain, nervous system, and consciousness are among these broad topics.

In the Theory of Everything, since everything in nature is made of indivisible atoms, we'll embark on another quantum journey together. Let's start with the heart: How does the heart continuously pump blood from birth to death? All multicellular organisms begin as a single cell. Humans and most species don't grow by dividing from a single cell but by assimilating external materials, increasing in volume and the number of elements, and progressing toward adulthood. Contrary to popular belief, they don't evolve from one species to another—they form directly. Take humans, for example: When a single cell begins to grow, the chemical reactions between the three main classes—elements, photons, and dark matter—form all our organs simultaneously in a very short time. The heart is one of these organs. The heart pumps blood through the vessels, but this pumping operates according to Pascal's principle. (Pascal's principle is defined as follows: "If pressure is applied externally to a liquid in a closed container, that pressure is transmitted equally to every point of the liquid and every part of the container." Here, the container is our blood vessels.) In this pumping process, the heart isn't a machine operating on action and reaction but more like a tool. Blood is composed of soft atoms (electrons), lightweight elements, and dark matter detached from its array (shaped like a ball bearing) surrounding the electrons and elements. The density of electrons in the blood gives it elasticity, much like a spring. To understand this elasticity, consider a simple experiment: Take a long, thin balloon and fill it with water. Squeeze the balloon from the middle. When you squeeze it, the balloon expands and resists the positive pressure you apply, creating an opposing positive pressure. Think of the heart in the same way: Your hand is the heart, and squeezing the balloon represents the closing of the heart valve. When the heart valve closes due to this reaction, the expanded soft atoms compress, creating an opposing force. During this compression, some lightweight elements bond together—what's called "dirty blood" is this element synthesis. The valve, unable to withstand the opposing force for long, opens, possibly because negative pressure forms in the uncompressed area, releasing the locked valve. The elements in the blood, freed from compression, accelerate toward the other side of the valve. For elasticity to persist, the number of lightweight elements in the blood must remain balanced by assimilating elements like oxygen through breathing. This process repeats until death. This is how the heart works at the quantum level. More simply: The heart compresses the blood in the vessels, creating resistance in the compressed blood; the opposing force from this resistance generates momentum in the blood, and these movements continue.

Nervous System: The internal structure of the nervous system, which spreads throughout our body like blood vessels, consists of molecular arrangements of elements formed between dark matter, with a high density of electrons around the elements. (You may recall this from the states of matter section.) The nervous system forms in an intermediate state between liquid and solid. Such formations enable the phenomenon of electricity. The nervous system works like squeezing the end of a wet sponge, causing water to come out the other end, but it's highly sensitive. When you touch something with your finger, the soft atom (electron) at the nerve ending is subjected to pressure. An electron is released from the other end of the nerve, initiating an algorithmic process in your brain, allowing you to understand what you've touched. When a painkiller is injected around a nerve, external elements are introduced, compressing that section of the nerve. This compression restricts the movement of electrons in the nerve, so you feel little or no pain.

Brain: The brain is a vital organ closely connected to the heart and nervous system. The nervous system feeds the brain with electrons (soft atoms), while the heart supplies it with elements (solid atom clusters). The nervous system doesn't continuously feed the brain with electrons, but the heart constantly supplies it with elements. Metaphorically, the brain is one of our hungriest organs, requiring constant nourishment. That's why, while we're awake, the brain is under a constant bombardment of solid and soft atoms. Soft atoms enter the brain in two ways: First, through the nervous system; second, through photons entering our eyes when we look at something (photon clusters of soft atoms turn into electrons the moment they enter our body) and through the sounds we hear with our ears. Elements reach the brain through blood vessels, pushed by the heart. Whether you dance, walk, or watch a movie, everything you do while awake happens thanks to the elements and photons delivered to your brain. Note: The cells of the nervous system are called neurons, but in the Theory of Everything, I chose to call the soft atoms around elements electrons to avoid conceptual confusion, so I'll continue using the term electrons.

Since all our organs form simultaneously, the relationship between the heart, nervous system, and brain can be likened to an engine idling. In this relationship, frequently repeated algorithmic processes become automated over time. We don't even need to strain our brain to walk, talk, swim, or ride a bike. The movements we frequently perform are the paths where algorithmic processes occur in our brain. The more a path is used, the easier it becomes to traverse. (The path is a metaphor, referring to the restricted areas where elements and/or electrons interact.) If we perform a physical action for the first time, solve a problem through thinking, or imagine a story, everything becomes more difficult in the brain. In such situations, the brain works to create new paths using its muscles. To use an analogy, the quantum events in our brain are like a giant, three-dimensional labyrinth made of plants and walls. (I believe this is currently called a synapse.) When we try a new movement or solve a problem for the first time, creating a path for algorithmic processes in the brain is like opening a door in the plant walls of the labyrinth—challenging quantum events occur in the brain. That's how our brain works while we're awake. But what about when we're asleep?

I'll explain the brain's working principle during sleep very simply. While awake, the brain is constantly bombarded with photons, electrons, and elements, but during sleep, the opposite happens: Elements and electrons flow from the brain to the body. When all the muscles in the brain relax, the ends of the blood vessels compressing the muscles open. Lightweight elements pass through these openings to be mixed into the blood by the heart. The number of beneficial lightweight elements in the blood increases. These elements are shared by other organs through the blood vessels, repairing the body. Similarly, electrons fill the cables of the nervous system, repairing the nerve cables worn out during the day. As the elements that nourish the brain while awake leave the brain, something else happens. We likened the brain to a three-dimensional labyrinth. As elements leave the brain, they pass through random paths, triggering the physical counterparts of information we gained while awake or long-forgotten, unused information. These random triggers cause dreams and nightmares. Thus, the entire mechanical mystery of the dream phenomenon is simply resolved.

100 HYPOTHESES

I've tried to explain the Theory of Everything starting from before the Big Bang. If I had attempted to explain this theory a thousand years ago, it might have been easier. Modern science, having been in use for many years, has increased the number of dogmatic individuals who blindly adhere to it. Additionally, although there hasn't been significant progress in modern science for about 50 years, the production of countless theses and articles has created substantial information clutter. This clutter may have made it harder for you to understand some of the topics in this book.

While explaining the Theory of Everything, I often had to refresh certain information. This information typically revolved around the internal structure of atoms, elements, photons, and dark matter. We live in a nature that inspires awe. Grasping that this nature is composed of a single fundamental particle—atoms—can be challenging for some. That's why I've decided to conclude the rest of the book with 100 hypotheses. Some of these hypotheses, and certain sections within them, will serve to refresh information.

In the Theory of Everything, describing every object and event in nature wouldn't fit in a single book—not even in an encyclopedia spanning millions of pages. We must also remember that some topics are very broad and comprehensive. While everything may not fit into a book, the mechanics of nature's operations can. The sole purpose of this book is to teach the micro-level workings of nature and the reactive behaviors of micro-particles. Once you learn these, I believe you'll be able to solve any natural mystery in seconds.

1. THE ATOM MODEL IN THE THEORY OF EVERYTHING

In this book, I've redefined terms used in current physics. The word "atom" is one of them. In the Theory of Everything, an atom is referred to as A.T.O.M., an abbreviation for Indivisible Singular Organic Matter.

The atom model in the Theory of Everything is fundamentally similar to John Dalton's atom model but describes the atom's properties more clearly. The fundamental principles of the Theory of Everything atom model are as follows:

- An atom has two extreme states: a solid, compact, raw, and smallest-volume particle state, and a soft, loose, and larger-volume state.
- An atom cannot be divided or destroyed. In its solid, raw form, it is the smallest unit in nature. Its small size makes it impossible to divide or create a device to divide it. Colliding two solid atoms in their singular form is also impossible because a solid atom expands when dislodged from its position.
- An atom that has expanded and increased in volume cannot be compressed back to its original state. This principle is the cause of the entropy we observe in nature. When a solid, small atom expands, its volume increases and it becomes soft. Due to its soft structure, pressure from two directions causes it to slip away. Compressing it from all directions fails because the materials applying pressure exert positive force on each other during the process.
- Atoms fill every space in nature.
- An atom, in its singular form, cannot wear out, be divided, or be destroyed. Such events only occur in clusters of atoms

at micro and macro scales. At the macro level, "destruction" is a metaphor.

2. DARK MATTER/ENERGY IN THE THEORY OF EVERYTHING

Dark matter is commonly described as "a substance that makes up a large portion of the universe's mass but cannot be directly observed." In the Theory of Everything, all mysteries of dark matter and energy have been unraveled and redefined. Like everything else in the quantum realm, dark matter cannot be directly observed, but its effects—from the quantum level to the macro scale—can be observed or perceived, just like other quantum particles.

Fundamental Principles of Dark Matter/Energy:

- The internal structure of dark matter and energy is the same, as both are composed of solid atoms. We call the largest clusters of solid atoms in regions dense with elements and photons "dark matter," and the largest symmetric clusters of solid atoms in areas where elements are absent and photon movement is sparse "dark energy." Dark matter and energy are the largest symmetric clusters of solid atoms.
- Dark matter is symmetric. Having formed before the universe, dark matter is symmetric and arranged in arrays. Dark matter within one array is super-symmetric with itself, while it is relatively symmetric compared to dark matter in another array.
- Under the right conditions, dark matter can break apart, detach from its array, or lose its symmetry. In violent natural events like the Big Bang, supernovae, or nuclear explosions, dark matter can fragment, with each fragment turning into asymmetric elements. High element and photon density can detach dark matter from its array. When completely surrounded by elements, dark matter

easily detaches from its array. All flowing liquids, like water, are composed of dark matter that has been detached from its array and surrounded by elements.

3. ELEMENTS IN THE THEORY OF EVERYTHING

The term "element" is one of the concepts I've adapted from current science and integrated into the Theory of Everything. In this theory, an element does not contradict experimental observations in current science but has a different internal structure as a model. Concepts like protons and neutrons do not exist in the Theory of Everything's element model.

Fundamental Principles of Elements:

- Elements are asymmetric. An element is an asymmetric cluster of solid atoms that emerges when symmetric dark matter breaks apart, remaining smaller than dark matter and trapped between dark matter clusters.
- Elements can break apart or combine through natural or artificial means. Due to their asymmetric structure, elements can stick together when their flat, broad surfaces come into contact. If their broad surfaces bond, they form strong molecular bonds; if other surfaces bond, the molecular bonds are weak. This property gives rise to countless different molecular bonds. Elements can be classified from small to large and exist as clusters of solid atoms in various asymmetric shapes. Therefore, there's no issue with using the current periodic table. Since elements come in countless asymmetric shapes, some of these shapes are symmetric with each other. When symmetric elements dominate, they can exclude asymmetric ones; when asymmetric elements dominate, they can exclude symmetric ones. This is why substances like oil, gas, gold, and silver are found in veins or pools in nature. It's also the reason behind the formation of living organisms' organs. In

current science, this is one of the causes of micro gravitational effects.

- The clustering of elements enables the formation of magnetic fields. Elements occupy space between symmetric dark matter, exerting positive force from the inside outward. This force causes the contact points in the arrays formed by dark matter to become more compact. The compressed dark matter bends the light passing around objects made of elements.

4. PHOTONS IN THE THEORY OF EVERYTHING

Photons are the fundamental triggering factor for everything, from the formation of life to the birth of the universe. A photon is a cluster of soft atoms, formed when raw, solid atoms expand and become soft. Being a cluster of soft atoms, a photon has a relatively light mass proportional to its volume. This structure of photons influences everything in nature. Photons form under specific conditions, rarely from dark matter clusters, but typically from the expansion of solid atoms vacuumed from elements.

Fundamental Principles of Photons:

- A photon's speed, trajectory, and behavior are determined by the pressure and balance exerted by dark matter. Photons are shapeless, liquid-like clusters that are asymmetric in number, not shape. Therefore, the number of soft atoms forming a photon determines its speed and behavior. As photons move through dark matter under pressure, they constantly lose or gain members during their journey. If a photon is subjected to balanced pressure from all directions, it can be trapped. When trapped in a large space, it forms objects like planetary or stellar cores; in a small space, it transitions into an electron form.
- Photons are the fundamental source of many energy forms. The shapeless and soft structure of the atoms forming a photon allows it to easily escape from its trapped location with the slightest change. When soft atoms in electron form push other soft atoms, flowing between elements, energies like electricity emerge. When electrons detach from an element and begin to move away through dark

matter, they take on the photon form, triggering light and heat energy. This is also the cause of entropy: The space left by a photon, which forms in an element cluster and departs through dark matter, is immediately filled by dark matter. This process continues until all the solid atoms in the element turn into soft atoms.

- Photons are always particulate in clusters.
  A photon is a cluster of soft atoms, making it inherently particulate. Soft atoms typically form through the vacuuming of solid atoms in an element. At the moment of vacuuming and while close to the element, they are in electron form. When they detach from the element and begin their journey, they transition into photon form. During the transition from electron to photon, a photon bombardment occurs. This bombardment, with photons of asymmetric sizes, enables the photon to be observed and detected by devices as both a particle and a wave. In reality, a photon is always particulate, but its production rate and the speed of light create a phenomenon where it appears as both a particle and a wave.
- At the quantum scale, a photon never moves in a straight line.
  This is due to the hexagonal, spherical structure of dark matter. Since photons travel only through the spaces between dark matter, their movement follows the shape of the path. Even when a photon appears to move in a straight line as light, at the quantum level, it travels while vibrating.
- Intense photon bombardments can create magnetic fields.
  Photons are lightweight clusters that take shape based on their environment. During events like the Big Bang,

supernovae, or stellar solar flares, when intense photon emissions occur between dark matter, the area occupied by photons causes dark matter to compress, forming a magnetic field.

5. PHOTOSYNTHESIS HYPOTHESIS

In the Theory of Everything, everything becomes simpler when you use your imagination. When the topics in this book are presented in a documentary format with animations, you'll be amazed at their simplicity. However, while reading the book, you'll need to rely on your imagination. To understand photosynthesis, it's helpful to know how a tree forms and functions at the quantum level. A living organism's genetic structure consists of asymmetric elements. For clarity, let's use examples in millimeters: For instance, a human's genetic structure might consist of elements with diameters ranging from 0 to 10 mm, while a tree's genetic structure might be made up of asymmetric elements with diameters ranging from 0 to 20 mm. In the Theory of Everything, this is how genetic structure is defined. In this example, if a human's formation begins with more 5 mm diameter elements than 6 mm ones, the human develops as male.

Turning to photosynthesis: A tree's intake part is its roots, and its output part is its leaves and fruits. The pressure exerted by dark matter on elements causes the smaller elements entering through the roots to be compressed and move upward through the larger elements. In simple terms, the tree takes in elements through its roots and expels them through its leaves and fruits. When photons from the Sun or another source strike the leaf, lightweight elements like oxygen that have accumulated in the leaf rise and disperse into the surroundings due to dark matter pressure. When the intense photon bombardment ends and it gets dark, the chaos caused by photons ceases, causing the lightweight elements on the leaf's surface to stick together, resulting in shrinkage. During this shrinkage, the elements that bond together transform into new elements, such as carbon dioxide.

What's called oxygen production is simply the separation of element clusters. In wavy seas, wave collisions can also separate oxygen from element clusters. Photosynthesis in water and plants occurs in this way.

6. VORTEX HYPOTHESIS

Vortices are phenomena we occasionally encounter in daily life. Every person has witnessed the formation of a vortex at least once. For example, if you turn on the faucet in a sink and close the drain, then open the drain, the water accumulated in the sink forms a hole in the center and spirals down the drain until it's gone. How does this happen? The liquid state of matter occurs when dark matter particles are surrounded by elements. Let's recall the bearing analogy: We considered the inner ring of the bearing as dark matter, the balls as elements, and ignored the outer ring. Dark matter detached from its array in this way allows the object to be liquid and flowable. Let's also remember that every object has a magnetic field. When we close the drain of the sink and fill it with enough water, a magnetic field and its lattice wall form on the liquid's surface. The moment we remove the plug, water begins to flow through the drain. At this point, dark matter starts to condense on the water's surface directly above the drain. The condensed dark matter, aided by the dark matter forming the magnetic wall, exerts a positive force from the inside outward, pushing the water elements aligned with the drain. The water, pressed from the center toward the edges, forms a vortex. In tornadoes rising from the ground to the sky, the opposite occurs: Dark matter in the air applies pressure from the outside inward, causing the elements in the center to rotate and rise in a funnel shape. The rotation of element clusters in vortices and tornadoes is due to the hexagonal, spherical structure of dark matter. A similar mechanism to the one observed in the sink also occurs in black holes. Since space is entirely covered with dark matter, countless vortices form in space. Black holes, as well as everything that rotates around its axis, including Earth and the Sun, are at the center of a vortex. Vortices are a key factor in preserving the momentum of Earth's ceaseless rotation.

7. MOMENTUM AND AERODYNAMICS IN SPACE

Momentum is a measure of an object's kinetic energy. Aerodynamics is the branch of science that studies the interaction of moving solid masses with air. The principle of momentum conservation states that in an isolated system, the total momentum remains constant over time—meaning that if no external forces act on the system, its total momentum stays the same.

In the Theory of Everything, an isolated system without elements consists solely of dark energy. However, when element clusters are introduced into such a system, the dark energy area transforms into a dark matter system. Since everything is made of atoms, every space is filled with atoms, and there is no truly isolated system in nature. Everything exists within everything; everything is one, and one is everything. Therefore, isolation is limited. A system can only be isolated up to the breaking point of the cage that encloses it. In space, countless clusters are relatively isolated based on the resistance of their magnetic field cages. Magnetic fields isolate element clusters to a limited extent. For example, Earth is isolated from space to a certain degree, which is why the elements in our atmosphere that sustain life don't escape into space.

Every space, whether inside or outside limited isolated systems, is covered with dark matter. From the womb, we carry certain dark matter particles within our cluster, and we retain them until we fully decompose after death. Once we completely decompose, the dark matter we've carried since birth rejoins symmetric dark matter arrays.

Every object has a magnetic field. When we move, the dark matter associated with us moves along with us. For instance, when we walk, our magnetic field cage wall slides between the dark matter arrays surrounding us. This phenomenon occurs in every moving object.

Now, let's address this question: "If every space is filled with dark matter, photons, and elements, how can we move so easily without remaining static, compact, or stationary?" To answer this, let's set up a Karmatan experiment, but this time, the surface of the setup is much larger and covered with air-blowing holes. In the center of an area covered with hexagonal rubber, we place asymmetric particles representing an element cluster object. Since the object is very compact, moving it will be difficult. However, when air—representing photons—is blown from the ground, the object cluster will move. This experiment is two-dimensional. Considering that our nature is three-dimensional, an object that moves slightly in our experimental setup would move exponentially more easily in a three-dimensional environment.

On Earth, momentum is short-lived because the elements in the air reduce its duration. In space, the mechanics of momentum are the same, but since elements are sparse, the duration of momentum conservation is much longer. Aerodynamics in space works similarly to that on Earth. Let's consider a meteor traveling through space: The elements forming the meteor, along with dark matter detached from its array and integrated into the element cluster, create a magnetic field around it with a positive force from the inside outward, forming a cage wall. The meteor's magnetic field cage slides through dark matter, but friction in space is much lower than on Earth. During momentum, the meteor exerts a positive force on the dark matter aligned with its direction of travel. The dark matter pushed to the meteor's sides gains kinetic energy due to its change in position and exerts a positive force back toward its original position, like a spring. In other words, it applies pressure from the sides until the meteor passes. After the meteor passes, the dark matter quickly returns to its original position, generating kinetic energy. This process also creates negative pressure, similar to the wind you feel on Earth when a car speeds past you.

In conclusion, aerodynamics exists in space and operates with the same mechanics as on Earth. Because elements are sparse in space, the kinetic energy generated in element clusters and dark matter arrays results in a much longer duration of momentum conservation.

8. WHERE DOES THE SUN GET ITS ENERGY AND HOW DOES IT SHINE FOR BILLIONS OF YEARS?

Let's start with a proposition that might make you say, "What's the connection?" It's impossible to shape an object made of elements into a perfect sphere. Space is composed of dark matter, so every area is covered with dark matter. The shape of dark matter resembles a sphere with hexagonal surfaces. Due to this structure, it's impossible to shape a cluster of elements into a perfect sphere; at the micro level, there will always be indentations and protrusions. Additionally, as the pressure from the arrays forming a sphere increases from the outside inward toward the center, the surface of the core—what we might call the center—experiences increased lateral pressure from the arrays. This can cause the sphere to take on a geoid shape or shift in and out of a geoid shape. Despite this, the smoothest spherical surfaces in space are the cores of stars like the Sun and planets. The cores of stars and planets are clusters of shapeless soft atoms. For this reason, the cores of stars and planets can temporarily take on a near-perfect spherical shape. The Sun's core is also a cluster of soft atoms. We've been calling soft atom clusters "photons." We can name the cores of stars "star photons" and the cores of planets "planet photons." Star and planet photons formed during the Big Bang and inflation.

During the Big Bang, when some dark matter broke apart and shattered, massive dust clouds made of elements emerged. The elements compressed the dark matter trapped between them toward the outer edges of the dust cloud. Countless dust clouds, each with a strength proportional to its mass, created magnetic fields around themselves. The cage walls of these magnetic fields reflected back most of the soft atoms in the photons escaping through the dust cloud, preventing them from passing through. The reflected photons, in the chaotic environment of the dust cloud, recombined under varying pressures to form massive photon clusters. These photon clusters would initiate the formation of planets and stars. In this chaotic environment, the largest photon clusters that couldn't withstand the pressure formed supernovae. Those that could withstand the pressure gained momentum, spinning around their own axes while instantly creating their own magnetic fields and vortices. The largest photon settled at the center, forming the main vortex. Being the largest, the central photon created the biggest vortex, drawing smaller photons into its influence. Each photon cluster within this massive vortex formed its own smaller vortices. Let's call this system of vortices within vortices our Solar System. Each photon at the center of a vortex began to draw the elements from the dust cloud—destined to become its core—into its magnetic cage wall. The elements, pushed by positive force and pulled by the winds created by the negative pressure described in aerodynamics, began clustering around the photon. The soft atoms forming the massive planet or star photon were so compact that no element could penetrate between them, forming tight molecular bonds around the photon's surface. Thus, our Sun was born as a star.

As soon as the Sun formed, it began radiating light. Despite its massive core, there weren't enough elements near its surface to completely cover it and block its light. In areas close to the surface, because the elements are asymmetric, solid atoms continuously turned into soft atoms, becoming electrons or escaping as light photons from the Sun. Since most of the elements covering the Sun's surface constantly produced soft atoms, this swelling environment compressed elements in other areas, creating temporary cold zones. (These cold zones on the Sun are referred to as "sunspots" in current science. Also, note that the dark spots on the Sun are similar in formation and structure to the tectonic plates on Earth.) This battle of forces on the Sun's surface resulted in a highly dynamic surface and atmosphere.

The elements on the Sun that serve as sources of heat and light will eventually be depleted. However, the massive explosions on the Sun—like those during the Big Bang and supernovae—break apart some dark matter particles, creating new elements. In other words, solar explosions fragment some dark matter, transforming it into elements. The solid atoms on the surface of these elements turn into soft atoms under negative pressure, becoming light photons and embarking on a journey away from the Sun. (Another note: One of the causes of solar explosions is the alignment of planets in the Solar System. For example, when Earth positions itself between Mars and the Sun, an increase in dark matter pressure—and a cooling effect—occurs on the Sun's surface, followed by an explosion.)

In conclusion, contrary to common belief, the Sun's energy source is the elements transformed from dark matter that breaks apart and fragments during violent explosions. By feeding on this process, the Sun can shine for billions of years.

9. WHY DOES HOT WATER FREEZE FASTER?

The phenomenon of hot water freezing faster is known as the Mpemba effect. This effect describes how hot water, under certain conditions, freezes faster than cold water. The reasons behind the Mpemba effect are not fully understood in current science.

Anyone well-versed in the Theory of Everything can logically explain all natural phenomena within seconds by adhering to the physical laws of this theory. The Mpemba effect is one such phenomenon.

First, we must remember a key principle: We need to use our imagination while grasping the simplicity of the quantum level. At the atomic level, there are no perceptual concepts like hot or cold. The atoms that make up an object are either in a solid state, bound to dark matter or element clusters, or in a soft state, existing as electrons on the surface of elements. At the macro level, the perception of hot and cold arises through a single natural or artificial process: Elements create the sensation of heat or cold based on their vibration and movement speed. If an element finds a wide space between dark matter—recall the fly-in-your-palm analogy—temperature increases; if it doesn't, temperature decreases. The availability of space for elements depends solely on the positive pressure exerted by dark matter from the outside inward. This is a battle of forces. If you subject an element cluster to intense photon bombardment from the outside, the incoming soft atoms will both change the element's position as they pass by and exert a positive force from the inside outward. This dominance of force increases the temperature. If the external force prevails, the temperature drops.

When we think of water, we think of matter in its liquid state. When lightweight elements with low density cluster together, they surround some dark matter particles like a bearing, detaching them from their array. When normal water is heated, it expands in all directions, pushing its surrounding magnetic field outward. This push causes the magnetic field, made of dark matter, to stretch like a spring. When we take hot water from its environment and place it in a refrigerator, the magnetic field of the hot water, which is stretched like a spring, has more kinetic energy compared to the magnetic field of water at a normal temperature. As a result, it exerts a positive force from the outside inward more quickly. This difference in positive force in the magnetic field, combined with other factors like evaporation, causes the hot water to freeze faster, resulting in the Mpemba effect.

10. DOPPLER EFFECT HYPOTHESIS

The Doppler effect describes the change in a wave's frequency and wavelength due to the relative motion between a moving source and an observer. This effect applies to both sound waves and electromagnetic waves (such as light).

In the Theory of Everything, as explained in the "Colors" section, solid atoms transition into a soft state and cluster together to form photons. A photon gains direction and speed based on the pressure exerted by dark matter. (This is similar to squeezing a wet bar of soap with your hand: The harder you squeeze, the farther the soap shoots out.) Since a photon is a cluster of soft atoms, it continuously loses elements during its journey. If it collides with a photon coming from another direction, it may gain elements. For example, if a photon starts its journey with 100 million elements, this number steadily decreases over time. This process gives rise to the Doppler effect.

Fundamental Principles:

- Approaching: When the source and observer move closer to each other, the frequency of the waves increases (wavelength shortens). In this case, the observer perceives a higher frequency (e.g., a higher-pitched sound or a bluer light color). A photon, when it first leaves the source, has its highest mass by volume, resulting in a higher speed and a shorter wavelength.
- Receding: When the source and observer move away from each other, the frequency of the waves decreases (wavelength lengthens). In this case, the observer perceives a lower frequency (e.g., a lower-pitched sound or a redder light color). As the photon loses elements during its

journey, the volume it occupies shrinks, causing the wavelength to lengthen. This phenomenon occurs similarly for both sound and light photons.

Examples:

- Sound Waves: The siren of an ambulance sounds higher-pitched (shriller) as it approaches you and lower-pitched (deeper) as it moves away. This is due to the siren's frequency changing based on the relative motion between you and the ambulance.
- Light Waves: In astronomy, if a star or galaxy is moving toward us, its light shifts toward blue (blueshift); if it's moving away, its light shifts toward red (redshift). This occurs because the frequency of the light changes depending on the star or galaxy's motion.

11. CASIMIR EFFECT

The Casimir effect (or Casimir force) is a significant phenomenon in quantum field theory and quantum electrodynamics. This effect arises due to the presence of zero-point energy of particles and fields, even in what is defined as a vacuum or absolute emptiness, resulting in an attractive or repulsive force. It was theoretically predicted by Hendrik Casimir and Dirk Polder in 1948. For example, a lizard can adhere to a surface using just the tip of a single toe, or two metal plates can stick together when placed close to each other.

In the Theory of Everything, the cause of the Casimir effect is quite simple and relates to the structure of magnetic fields, as I've mentioned several times before. Every visible object has a magnetic field and a magnetic field cage wall. A lizard can adhere to a surface because the magnetic field cage wall at the tip of its toe merges with the dark matter in the magnetic field cage wall of the surface it clings to. Factors such as the molecular structure of the toe and the lizard's weight also play a role in this ability. Similarly, when two flat metal plates are brought close together, their elements don't initially touch. First, their magnetic fields make contact, followed by their magnetic field cage walls. These cage walls are a compressed form of dark matter. When dark matter in the cage walls merges, it becomes even more compressed, combining two forces into a stronger one. In current science, terms like "absolute emptiness" actually refer to dark matter that cannot be directly observed, but scientists are unaware of this.

12. PASCAL'S PRINCIPLE

Pascal's principle is a fundamental concept in pressure physics and can be summarized as follows: In a closed liquid system, pressure applied at one point is transmitted equally to every point in the liquid. This principle was discovered by the French scientist Blaise Pascal and forms the foundation of how hydraulic systems work. For example, in a hydraulic lift, a force applied to a small piston is transmitted through the liquid, creating a larger force at a larger piston, allowing heavy loads to be lifted with minimal effort.

The Theory of Everything enables you to easily unravel the internal structure and mechanical mysteries of everything in nature. Pressure physics underlies every event in nature. A change in pressure at any point in liquids and gases can spread through chain reactions to a wider area, creating dynamic and chaotic environments. We can observe this in all meteorological events. While Pascal's principle focuses on liquid pressure in closed systems, let me remind you that there is no completely isolated system in nature.

The liquid state of matter arises when dark matter, detached from its array, is surrounded by elements. Let's recall the bearing analogy: The hub of the bearing is dark matter, and the balls around it are elements. When the balls—representing elements—are lightweight, the liquid becomes superfluid; when they are heavy, the liquid is less fluid. Dark matter particles surrounded by lightweight elements have more electrons, while those surrounded by heavy elements have fewer.

In a hydraulic lift, the structure of the elements forming the liquid increases the power level through chain reactions. This process is similar to the wave function created in still water, but since it occurs in a closed system, the wave function happens between elements. It's like a small wave starting at one point that grows and increases in strength as it progresses. Dark matter plays a significant role in this process. In a hydraulic lift, some dark matter particles separate from the liquid and join dark matter arrays. As these integrations continue, the lifting force increases exponentially. When the lifting process ends, the system is relaxed, and the dark matter in the arrays rejoins the liquid.

13. PLANET FORMATION

All the planets in our Solar System began forming simultaneously with the Sun. During the Big Bang, some dark matter fragmented, giving rise to countless elements and photons. The dust cloud formed by these elements immediately created a magnetic field and a magnetic field cage wall. While a small portion of the photons hitting the magnetic field cage wall passed through, most were reflected back into the dust cloud. In the chaotic environment, photons that found a balanced space aligned at proportional intervals, becoming massive photons that would later form planetary cores. These massive photons, while spinning within the Sun's vortex, generated their own magnetic fields and vortices, gathering all surrounding elements around their cores.

One of the most critical factors determining a planet's outer surface structure is the diameter of the photon forming its core. If the photon's diameter is too large, it cannot withstand the pressure of dark matter and explodes as a supernova. If the photon's diameter is too small, it gradually disintegrates and disappears over time. You can better visualize this process through an image at https://www.eso.org/public/images/eso2008a/ (Accessed: May 10, 2025). The image shows two photon groups bouncing off the magnetic field cage wall of dark matter and regrouping within the dust cloud. When photons are trapped between dark matter and become too large in volume, they create their own magnetic field. This magnetic field causes the photons to orbit each other, much like two spherical magnets. During this rotation, the photons and surrounding dark matter form a vortex. Within this vortex, the two large photon groups eventually merge at the center. If the resulting photon bubble from this merger is small enough to withstand dark matter pressure, a new star or planetary core is formed. However, if the new photon bubble is too large, it cannot withstand dark matter pressure and results in a supernova explosion.

Let me illustrate the importance of core diameter with two examples:

Example 1: Assume that the photons forming the cores of Mercury, Venus, and Earth have the same diameter. Mercury, upon formation, joined the Sun's vortex and began collecting elements within its own vortex. However, being very close to the Sun, most of its elements were pulled away by the Sun. Thus, despite having a core with the same diameter as Earth, Mercury is smaller than Earth. Venus, located between Earth and Mercury, collected all the elements within its vortex around itself. However, because it was closer to the center of the dust cloud, it gathered fewer elements along its rotational axis compared to Earth, transforming into a planet with a dense atmosphere and a greenhouse effect. Earth, with the elements it gathered, became a planet suitable for life. Mars, on the other hand, has a core photon with a much smaller diameter than those of Earth, Venus, and Mercury, but its angular rotational diameter is larger due to its greater distance from the Sun. As a result, Mars couldn't collect many elements, leading to the formation of the asteroid belt between Mars and Jupiter.

Example 2: Jupiter, Saturn, Uranus, and Neptune have the largest core photons after the Sun. Due to their angular distances from the center, they are exponentially more massive than the Sun in terms of element count. Because of their large cores, these planets are gas giants. While the Sun's volume is exponentially greater than Uranus's, its element count is significantly less than Uranus's.

With these two examples, I aimed to demonstrate the impact of core diameter and the elements gathered around it on the structure of planets and stars.

In stars and planets, the distance between two cores is as important as the core diameter. The cores of stars and planets are regions free of dark matter—in other words, they are pure clusters of soft atoms. However, since they are situated between dark matter, a planet's magnetic field diameter and strength are determined and formed at the core. The magnetic fields generated at the core act as a barrier, preventing the Moon from falling onto Earth and the planets from colliding with the Sun. However, in the universe, cores are not always aligned at proportional distances. In very rare cases, two stars or planets that were initially at a proportional distance may come too close to each other due to environmental factors. When this happens, their magnetic field barriers are breached, and the two stars or planets begin orbiting each other. Their rotational speed increases as the cores get closer, similar to the motion of two magnetic balls. When these orbiting cores collide, they create a different type of supernova. A Type 1 supernova results from the collision of two cores, while a Type 2 supernova occurs when a massive photon, due to its large volume, cannot withstand dark matter pressure and explodes. The reason for supernovae is that the photon forming the core of a planet or star succumbs to dark matter pressure due to its size. This is akin to a large air bubble forming in water or a bubble growing underwater that bursts under pressure.

14. EARTH'S INTERNAL STRUCTURE AND EARTHQUAKES

The Earth's core diameter, the number of elements it has gathered, and the frequency of photon bombardment from the Sun have come together to create a perfect environment for life.

We cannot directly observe the Earth's internal structure by cutting it open like an apple. Instead, we can only estimate it by adhering to the laws of a given theory. For instance, the models of the Earth's internal structure you've seen so far have been developed based on a specific theory. In the Theory of Everything, we will follow the same approach, modeling the Earth's internal structure by adhering to its established physical laws.

Let's first recall some of these physical laws: Everything in nature is made of atoms. Atoms cannot be created from nothing nor destroyed when they exist; they cannot be divided, fragmented, or worn down. In their raw form, atoms are compact, small in mass, and solid. In their expanded form, they are soft and larger in volume and mass. Once an atom expands from its raw state to its soft state, it cannot be returned to its original solid state. (This is because, when an atom expands, equal pressure cannot be applied from all directions to compress it back to its solid state; this is what makes entropy possible.)

In the Theory of Everything, the only fundamental force is the positive force, which is the resistance exerted by an atom from the inside outward. This single fundamental force is the primary factor in the formation of all energy types and phenomena known in modern science. Gravity and gravitational attraction do not exist at quantum or macro scales; I use these terms only to describe certain phenomena. No macro clusters that are not in direct contact can attract each other.

At the quantum level, there is no concept of hot or cold. All energies, including heat energy, arise relative to the speed of atomic motion and the trajectory of photons. If the pressure exerted by dark matter on elements dominates, the space for elements to vibrate narrows, leading to cooling at the macro level. If elements find a wider space to vibrate—due to their asymmetric structure or photon bombardment—they temporarily overcome dark matter pressure, resulting in heating. Additionally, when the force of elements and photons surpasses dark matter pressure, dark matter compresses, affecting the strength of the magnetic field. When the magnetic field's strength increases, cooling occurs in another region. In other words, when a hot environment forms in one area, a cold environment forms elsewhere. This phenomenon is more commonly observed on spherical surfaces, with countless examples like Mercury and Earth.

With this refresher, we can model the Earth's internal structure as follows, dividing it into layers from the inside out:

- Core: A giant photon composed of soft atoms.
- Layer 1: As the Earth's core formed, it was immediately surrounded by symmetric dark matter, creating a thin dark matter layer around the core.
- Layer 2: While the Earth, with its own vortex, was caught in the Sun's vortex and spun within the dust cloud, the asymmetric elements surrounding it clustered from the inside outward, forming vein-like structures. Thus, Layer 2 forms layers resembling tree roots that extend up to the Earth's surface. In modern science, these layers are called tectonic plates. Layer 2 extends to the surface, with elements arranged in molecular patterns that transition

from compact to loose as they move outward. Consequently, the region around the Earth's core is cold.
- Layer 3: The Earth's surface, which we can think of as its skin or clothing, includes the crust, oceans, and the immediate subsurface regions. The Earth's surface is as we know it.

Since the roots of tectonic plates extend down to Layer 1, earthquakes are neither frequent nor large in scale. Additionally, due to the scarcity of volcanoes, the Earth's surface is generally not mountainous. Volcanic eruptions and earthquakes are related to factors such as the Moon's orbit and distance, the geometric alignment of planets, axial shifts during seasonal transitions, and solar flares from the Sun. These are the primary factors that trigger volcanic eruptions and earthquakes.

Let me provide a brief example: If tectonic plates move in opposite directions due to factors like the Moon, the geometric alignment of planets, axial shifts during seasonal transitions, or solar flares, a massive negative pressure forms beneath the Earth's crust. This negative pressure causes elements to vibrate and produce electrons. When solid and stationary elements begin producing electrons, heat energy increases. In a closed environment, this is similar to the phenomenon of a heated liquid releasing steam, as seen in a pressure cooker or a teapot. When this phenomenon occurs, the ground turns into lava and erupts through a volcano's mouth. If there is no nearby volcano to erupt, an earthquake occurs. A similar event happens when two tectonic plates exert positive force toward each other, but in this case, the earthquake or volcanic eruption occurs farther away. This is because when a tectonic plate applies positive force in one direction, a negative force forms in the opposite direction at the same time, or vice versa.

15. THE MOON

In the Theory of Everything, we continue to address some natural mysteries—whether solved, unsolved, or incorrectly solved. The Moon is one such mystery. While we know a lot about the Moon, some of this knowledge is incorrect or incomplete according to the Theory of Everything.

The formation of all planetary moons, including the Moon, is similar to that of planets. However, the photons forming the cores of moons are much smaller in diameter compared to those forming planetary cores. Because their cores are so small, some moons may have lost their momentum over time or disintegrated and dissipated.

The Earth's moon, the Moon, is one such example, which is why it does not rotate on its own axis. Just as the Earth orbits within the Sun's vortex, the Moon orbits the Earth, caught in the vortex created by the Earth's magnetic field. You can observe this with a simple experiment: Fill a sink with water and plug the drain. Take two light balls—one large and one small. Remove the plug, and immediately place the large ball at the center and the small ball near the edge. You'll notice that the large ball at the center rotates on its own axis, while the small ball orbits the large ball but does not rotate on its own axis. Just like in this experiment, the Moon does not rotate on its own axis. That's why, when viewed from Earth, we always see the same face of the Moon.

As the Moon orbits the Earth, it traces a zigzag path above and below the equatorial plane while simultaneously moving closer to and farther from the Earth in a cyclical pattern. During this orbit, it influences phenomena like tides and the flow of gases in the atmosphere. In tides, the Moon affects the water with a positive force, similar to the interaction between two magnets. When the Moon approaches the Earth, the dark matter in the space between them compresses. This compressed dark matter impacts the atmosphere, tectonic plates, and ocean water, triggering phenomena such as hurricanes, earthquakes, and tides. When the Moon moves farther away, it creates negative pressure and force in the dark matter. Thus, the Moon plays a key role in making the Earth's atmosphere dynamic, serving as one of the primary factors in atmospheric events depending on its position and motion.

16. EFFECTS OF LIGHT PHOTONS AT THE MACRO SCALE AND THE BEHAVIOR OF LIGHT

Since photons are clusters of soft atoms, classifying them based on the number of elements in the cluster or their diameter yields more accurate results. In modern science, however, photons are classified according to their wavelength. The primary factor determining the wavelength of light is the number of soft atoms in the photon and the degree of positive pressure exerted by dark matter on the light. The pressure exerted by dark matter and the volume occupied by the photon can vary depending on the number of soft atoms constituting the photon. This prevents the photon from maintaining a constant speed. For example, environments where dark matter pressure is strong are cold. Since a photon accelerates in response to the pressure it experiences, it travels faster in cold environments. This rule also applies in space. In space, the trajectory and speed of light are shaped by the pressure exerted by dark matter. Light traveling through the void between two galaxies might reach speeds 2–3 times its usual rate, as the pressure exerted by dark matter in element-free regions dominated by dark energy can reach immense levels. In areas with dark energy, dark matter can be so compressed that a photon may lose fewer elements during its journey. However, if dark matter becomes even more compressed, phenomena like black holes may prevent photons from passing through, reflecting all their elements back.

These and similar reasons cause light to give rise to countless phenomena in space at both macro and micro scales. Since photons are particle-like clusters, as explained in the Doppler effect, they can lose elements or gain elements lost by another photon. As photons move farther from their source, they lose elements, and by the time they reach our eyes or devices, they lose the visual information they carry. This is why we cannot see a bird in a tree on a mountain 10 km away. The same principle applies in space. Even if we built a telescope the size of Earth, the result would be the same, because the photon no longer retains the same number of elements it had when it first left its source. Magnetic fields can distort the path of light, and if the cage walls of magnetic fields are very strong, they can reflect a large portion of the light back like a mirror. Phenomena like black holes or the magnetic field cage walls of a galaxy can act as mirrors in space. Many of the countless stars we see in the sky might actually be reflections. For instance, the North Star, also known as Polaris, could be a reflection of the light emitted by our Sun, bouncing off the magnetic field cage wall of our galaxy, the Milky Way.

17. SEASONS

The Sun is vastly larger than the Earth. Since the Sun's core is the largest photon in the Solar System, its magnetic field is extremely powerful, creating the largest vortex in the system. The Earth, influenced by this vortex, orbits both the Sun and its own axis.

As the Earth orbits the Sun, relative to the Sun's equatorial plane, it follows a path of upward and downward zigzags while simultaneously moving closer to and farther from the Sun. The photons emitted by the Sun are particularly dense along its equatorial plane. This density occurs because, due to the Sun's massive size and spherical shape, the closest point to the Earth is along the Sun's equator. Thus, the light from the Sun affects the Earth like the forceful flow of a waterfall at its initial impact point, preventing the Earth from deviating from its orbital alignment. In other words, the Earth does not rise to the Sun's upper plane. You can better understand this by examining other planets. When the Earth moves above the Sun's equatorial plane, the southern hemisphere receives more direct sunlight, experiencing summer. When it dips below this plane, summer occurs in the northern hemisphere.

The formation of seasons is not related to the Earth's orbit around the Sun but solely to its axial tilt. You can observe this by conducting an experiment with water or a laser setup. Here's a simple experiment: Gather a hemispherical strainer, a round fruit like a plum or cherry, and a sink. Turn on the faucet and let the water flow. Place the fruit inside the strainer and position it under the flowing water. When the water flows directly onto the center of the fruit, the magnetic field of the water will trap the fruit. The trapped fruit will start rotating on its own axis and wobble due to the water's pressure, but it will not escape. The relationship between the Sun and the Earth is similar. The photon bombardment from the Sun, much like the flowing water from the faucet, traps the Earth's entire structure, including its magnetic field. (The magnetic field created by this photon bombardment can be likened to electromagnetism in modern physics.) As a result, neither the Earth nor any other planet in the Solar System escapes or deviates from its orbit relative to the Sun's plane, which is why the Earth has been orbiting the Sun for billions of years.

Furthermore, the intense photon bombardment the Earth experiences while aligned with the Sun's equatorial plane strengthens the Earth's magnetic field due to the area covered by the photons. The dark matter within this strengthened magnetic field exerts pressure on the Earth from the poles. This pressure varies depending on the Earth's position in its alignment. In the hemisphere experiencing summer, the Earth's surface expands and swells, while in the hemisphere experiencing winter, the surface contracts. Consequently, the Earth's shape slightly resembles a heart symbol: the summer side is "fatter," and the winter side is "thinner." This causes distances in the northern and southern hemispheres to vary with the seasons. This hypothesis is testable.

18. NUCLEAR BOMBS AND FUEL

In nature, including the Big Bang, one of the unchanging principles of quantum mechanics is the mechanism of natural or artificial energy production. At the quantum level, energy is released only when atoms transition from a solid state to a soft state. As you may recall from the Big Bang, in a space composed entirely of symmetric dark matter, the symmetry in the central dark matter arrays was disrupted, causing solid atoms to expand and ignite the first spark. All forms of energy you know—fire, lasers, light, electrons—are simply the transition of solid atoms to a soft state. Whether you light a match or detonate a nuclear bomb, the same process occurs, though the scale differs.

Nuclear bombs come in two types: fission (atomic) and fusion (hydrogen) bombs. First, let me clarify that in nuclear bombs, it is not atoms but elements that are split. In both fission and fusion bombs, elements are compressed purely between dark matter. This compression is achieved by rapidly propelling the elements, aligning them between dark matter. The symmetry of the compressed elements is crucial. To make this clearer: when elements the size of a cubic watermelon are compressed, a fission bomb is created; when elements the size of a corn kernel are compressed, a fusion bomb is produced. Elements are compressed between dark matter, but there is a limit to this process; the pressure of dark matter does not allow compression over a very large area.

Let's assume we've prepared our bomb. When a fission or fusion bomb detonates, countless soft atoms (electrons) are generated from the compressed elements. So many electrons flood the environment that the elements in the nuclear bomb find themselves in a photon pool. As I explained in the Big Bang, the element clusters, which are solid atoms, slide through the soft atoms and reach the speed of light. Elements reaching the speed of light collide with dark matter arrays due to their gained momentum. The collision of two solid entities creates another photon pool, dislodging some dark matter from their arrays and propelling them at the speed of light as well. This initiates chain reactions. The destructive power of a nuclear bomb stems from its ability to fragment some dark matter. The fragmentation of dark matter is similar to the Sun finding external fuel for energy production.

In nuclear fuel rods, however, the arrangement of elements is neither pure nor perfectly symmetric. Fuel rods contain small elements and dark matter between the larger elements, which is why the chain reaction is slower.

The immense destructive power of nuclear bombs is nothing compared to supernovae. While nuclear bombs are pure element clusters compressed between dark matter, supernovae are massive photons—essentially enormous clusters of soft atoms. Supernovae, being voluminous but lightweight clusters of soft atoms, are exponentially more resistant to the pressure of dark matter. I hope no one attempts to create a photon bomb in the future. Although I'd like to say, "Creating a photon bomb or a mini-supernova on Earth is impossible due to the structural properties of soft atoms," I hesitate to call it impossible, given the ingenuity of human intelligence and the rapid pace of technological advancement.

19. NEUTRON STARS

You're already familiar with neutron stars, as I've mentioned them before. However, since they are significantly misunderstood and incompletely defined in modern science, I wanted to dedicate a separate section to discussing neutron stars in detail.

Neutron stars are massive clusters of soft atoms. To put it simply, if the diameter of a photon forming a laser beam you've seen is 1 mm, the diameter of a photon referred to as a neutron star might be around 20 km. Neutron stars form when light photons escaping the magnetic field cage wall of a galaxy or its surrounding dust cloud merge together. The lifespan of neutron stars is longer than that of stars and planets because they serve as the cores of stars and planets. Depending on the density of elements in the surrounding dust cloud, they can evolve into stars, planets, pulsars, or magnetars.

Since neutron stars occupy space among dark matter, they immediately create a magnetic field and a vortex upon formation. Their magnetic fields are powerful, but the mechanics of phenomena labeled as gravitational pull or gravity are misunderstood in modern science. For example, you might have come across the hypothesis that "neutron stars are extremely dense; a teaspoon of neutron star material could weigh billions of tons on Earth." However, according to the Theory of Everything, this is not possible. On Earth, a teaspoon of neutron star material would only produce a lightning bolt—similar to the lightning and thunder you see during stormy weather.

20. MYSTERIOUS LIGHT ORBS

Mysterious light orbs are a phenomenon reported across many cultures and regions worldwide, yet they remain poorly understood. These light orbs are typically seen in the sky or on the ground and can vary in shape, color, and size. Such phenomena can be examined in several categories:

Scientific and Natural Explanations:

- St. Elmo's Fire: This phenomenon appears as a blue or purple glow, often observed on ship masts, airplane wings, and sometimes tall structures during electrically charged weather conditions. It typically occurs during stormy weather.
- Ball Lightning: A rare atmospheric event where a bright, usually spherical ball of light is observed floating in the air during a storm. Ball lightning can last from a few seconds to a few minutes and may appear in various colors.
- Phosphorescence: Certain organic materials or microorganisms can produce bioluminescence, glowing at night. In particular, plankton in the ocean can create such glowing effects.
- Hessdalen Lights: Observed in Norway's Hessdalen Valley, these light orbs have been scientifically studied. Various hypotheses have been proposed, but no definitive explanation has been found.
- Min Min Lights: Reported in Australia's Outback, these lights are bright orbs typically seen at night, with no identifiable source.

According to the Theory of Everything, such phenomena are entirely possible. Every light you see is formed by solid atoms expanding and transforming into soft atoms. Soft atoms behave in accordance with the pressure and balance of dark matter. Asymmetric photons can give rise to these rare phenomena.

21. CANCER

In the Theory of Everything, unraveling the mysteries of nature at both quantum and macro levels is quite simple and straightforward. According to the theory, atoms exist in two extreme states: solid and soft. Despite these two states, the fact that solid atoms form countless symmetric and asymmetric clusters, and soft atoms form countless numerically asymmetric clusters, results in an infinite number of combinations. This is the fundamental reason for the diversity in nature.

In modern science, cancer is defined as a group of diseases caused by the uncontrolled growth and division of cells. Normally, cells grow, divide, and die in a regulated manner after a certain period. However, cancer cells disrupt this normal process, continuing to grow and divide uncontrollably, eventually forming tumors (clusters of cells).

Modern science has developed based on existing theories. However, let me first clarify: According to the Theory of Everything, it is not possible for cells to multiply by dividing and form large structures like humans. A cell or cluster grows only when it is nourished externally. When elements cluster together, they immediately trigger the formation of a magnetic field and a magnetic field cage by dark matter. If the elements forming a cell cannot resist the pressure of dark matter with sufficient force, the cell divides. This means that cells divide because they cannot achieve a molecular arrangement capable of resisting dark matter pressure. I should also note that the frequency strength of photons can disrupt the magnetic barrier of a cell. If a cell divides immediately after forming in its natural environment, a large living organism can never develop from that cell.

In conclusion, humans are not formed from dividing cells. A human develops from a single cell that, under the appropriate photon frequency, assimilates external elements. The human genetic structure is composed of an element pool of asymmetric light elements with diameters between 0-10 millimeters. If elements with diameters of 15-20 millimeters or larger suddenly enter this element pool and cannot be expelled by the digestive system, cancer develops. This is because elements larger than 0-10 millimeters are not compatible with your genetic structure. To put it another way: If a cell began forming in an element pool of 0-20 millimeters, it would not develop into a human but into a tree or another type of organism.

The cause of cancer lies in asymmetric, large elements—relative to your genetic structure—that either enter the body through pathways outside the digestive system or infiltrate your organs after entering the digestive system without being expelled. These large elements gather beneficial elements around themselves, forming external clusters in your body. These clusters make you ill, leading to cancer symptoms. Additionally, exposure to intense photon bombardment or strong magnetic fields can also cause disruptions in your genetic structure.

I have briefly explained cancer according to the Theory of Everything. As a scientific discipline based on this theory develops, treating numerous diseases, including cancer, could become much easier.

22. WHY DO LIVING THINGS DIE?

Answering this question requires understanding how living things are formed in the Theory of Everything. Some of you might have already answered it, but let me provide a different perspective. According to the theory, everything in nature—living or nonliving—is composed of elements, which are asymmetric clusters of solid atoms. These elements are the fragmented forms of some symmetric dark matter, existing among mostly unfragmented symmetric dark matter. Soft atoms near an element form clusters as electrons, while those farther away form clusters as photons. Since light elements can easily stick together, they form viruses. Viruses are already present on all planets in the universe. All living organisms begin to form when viruses are exposed to the appropriate photon frequency. The process that starts with the dissociation of viruses, progressing from a single cell to adulthood and eventually death, is the evolutionary line between life and death. At the core of the greatest struggles along this line is the battle between the external positive pressure exerted by dark matter and the internal outward resistance of a living organism, which has incorporated some dark matter into its cluster. As a living being, you are literally—not metaphorically—bearing the weight of the entire universe. However, the Earth's atmosphere, the Solar System, and the galaxy assist you in this struggle. How? These structures create negative spaces (meaning the looseness between dark matter) within their magnetic fields to provide support. This is why a fish cannot survive long on land, and a human cannot survive long in space. A human left in space without a protective suit would instantly freeze and die because the dark matter pressure in space is exponentially stronger than on Earth. On Earth, if dark matter pressure begins to dominate, the dying process starts, and eventually, you die.

If you've understood the concept of pressure, let's continue from the single-cell stage. The moment a virus dissociates and becomes a single cell, it occupies space among dark matter, creating a magnetic field and forming its own magnetic field cage (a cell barrier). This magnetic field cage exerts a resistance force against dark matter pressure. If this resistance breaks, the cell dissociates and reverts to a virus. The only way to prevent it from reverting to a virus is through nourishment. When a cell assimilates external elements, it occupies more space in terms of mass and volume, strengthening its magnetic field. As the cell finds more nourishment, it enters a rapid growth phase. Let's assume the initial virus forming this cell has a genetic (DNA) structure composed of asymmetric elements with diameters of 0-10 millimeters, as mentioned earlier. As it enters the growth phase, all its organs form instantly. The cell grows as it is nourished by elements, eventually reaching adulthood. In adulthood, the external pressure from dark matter is balanced by the cell's internal outward positive force (resistance). This is why young adult humans no longer grow taller or larger; the growth process has stopped. The elements assimilated during nourishment have been absorbed by the organs, meaning the heart, lungs, brain, and other organs have also grown. Each organ has its own magnetic field and barrier because organs have symmetric molecular arrangements. For example, the element arrangement of the heart differs from that of the lungs, which is why organs reach a growth capacity limit.

When organs stop growing, they must maintain balance. The struggle between the body's internal outward resistance and the external pressure from dark matter never ceases. This balanced struggle continues until the onset of aging. During the dynamic process from adulthood to old age, organs begin to wear out. Some organs start to lose their resistance to dark matter pressure and begin to wither. As all organs decline, death approaches, and dark matter overcomes the body's resistance entirely. When a human's resistance is completely broken, they die, dissociating and leaving behind a legacy of viruses in nature. Some of these dissociated viruses, in the initial stage after death, can form new cells depending on environmental factors, resulting in organisms like worms, mold, or fungi.

In conclusion, living things die because they lose the battle against the pressure exerted by dark matter. The lifespan of living organisms depends on the size of their elements and their molecular arrangement. For example, most trees outlive humans because their genetic structure primarily consists of elements with diameters between 10-20 millimeters. Some turtles can live for 200 years because their genetic arrangement results in a shell that acts as a thick skin. This shell creates a strong magnetic field cage, allowing them to withstand dark matter pressure for a longer time. The lifespan of everything—whether living or nonliving, composed of elements—depends on this mechanism. Be it a star, a rock, an insect, or a human, if dark matter prevails, you either perish or die. This mechanism also explains why humans gain or lose weight. If you overnourish yourself, your internal outward positive force increases, and you gain weight. If you don't nourish yourself, you lose weight, become ill, and eventually die.

23. MAGNETIC FIELD

In modern science, a magnetic field is defined as "a vector field that arises from the motion of electric charges and demonstrates the effect of magnetic forces." According to this definition, the magnetic field is used to describe and analyze magnetic forces, with its sources typically being magnets, electric currents, and changing electric fields. However, based on the Theory of Everything, this definition is incorrect and incomplete. I continue to correct the inaccurate or incomplete definitions of natural phenomena based on existing theories.

In the Theory of Everything, a magnetic field is an area formed by the compression of dark matter, resulting from the positive force or resistance exerted outward by elements and photons that occupy space between or within symmetric dark matter. This is similar to what we observe in the Karmatan experiment. In an area filled with symmetric particles, when the symmetry of some symmetric particles is disrupted, these particles form new asymmetric particles (elements). These newly formed asymmetric particles exert an outward force, creating vectorial force lines within the arrays of symmetric dark matter. The vectorial force lines formed between dark matter are the reason for the pole directions of a magnet. The sources of magnetic fields are solely elements and photons, while the magnetic field itself consists of dark matter.

Every element cluster occupying space between dark matter can generate a magnetic field as an object. Objects with element arrangements like those of a magnet, as well as temporary forces externally applied by photons or electrons, can create vectorial force lines. Magnetic fields are formed when elements and photons exert an outward force, causing dark matter to compress each other. This compression is generally balanced and proportional. The imaginary vectorial lines formed in this balanced proportion are why countless objects—like apples, humans, trees, iron, paper, or stones—do not exhibit magnetic properties. However, in objects that do exhibit magnetic properties, the vectorial fields are different; dark matter is more compressed on one side of the object and less compressed on the other. In the Karmatan experiment setup, when we wind the symmetric structures we assume to be dark matter in one direction like a spool, the resulting effect better explains the magnetic field with magnetic properties.

The element arrangements in natural magnets compress the dark matter on one side more than the other. When the compressed sides of two magnets approach each other, the dark matter causes the magnets to repel each other. In all other cases, they attract each other.

24. INDUCTION HEATING AND ELECTROMAGNETS

In magnets, the vectorial lines of the magnetic field are formed based on the molecular arrangement of elements. This formation causes dark matter in one direction to become highly compressed while dark matter in another direction remains less compressed. Imagine dark matter as a rope extending from the object, with the elements inside the magnet acting like a spool. When we apply electricity to a suitable material (an induction coil) with a circular element structure, the positive force of the electrons externally added to the cluster changes the position of the elements. As a result, the elements, which we imagine as a spool, pull the dark matter arranged like a rope toward themselves. When this process occurs across countless elements, a magnetic field with negative pressure emerges in the empty space at the center of the induction coil. As we learned earlier, heat energy increases only in areas with negative pressure, in the absence of external photon bombardment. In the empty space at the center of the induction coil's circular structure, dark matter shifts into a very loose position. When we place a metal in this area of loose dark matter, the elements forming the metal initiate rapid electron emission between the loose dark matter, generating heat energy. This is the quantum mechanics of induction heating.

An electromagnet works in the exact opposite way to induction heating. While induction heating creates negative pressure at the center of the coil, an electromagnet creates positive pressure at the center. This positive pressure enhances the magnetic field effect of the iron, causing dark matter to compress more with positive force. Any metal entering the area of compressed dark matter created by the electromagnet is attracted to this field.

25. AURA

An aura is generally defined as an invisible energy field believed to surround a person or object. In spiritual and esoteric traditions, it is thought to reflect an individual's physical, emotional, mental, and spiritual state. Auras are often described in various colors, with each color believed to carry a different meaning.

In the Theory of Everything, an aura is a magnetic field composed of dark matter. Developing a hypothesis about auras can help us better understand the nature and interactions of dark matter. As I previously categorized, everything in the universe falls into three classes: elements (composed of atoms), dark matter, and clusters of photons/electrons. This includes humans and all objects. These clusters cause the surrounding symmetric dark matter to envelop them with the potential energy of dark matter, as described in the definition of a magnetic field or aura.

Every object emits radiation into the dark matter that forms it, depending on the structural arrangement of its elements and electrons. This radiation passes through the object's initial magnetic field (the primary aura layer), creating energy fluctuations as it does so. At the quantum level, these energy fluctuations arise from the interaction of vibrations caused by the asymmetric arrangement of elements with dark matter. For example, the radiation emitted by a calm person produces lower-frequency vibrations, while an excited person's elements generate higher-frequency vibrations. These vibrations create varying energy densities in the dark matter layers of the magnetic field (aura), which may historically have been associated with colors. For instance, a calm person's aura might be linked to blue tones, while an excited person's aura might be associated with red tones, reflecting the differing intensities of energy fluctuations in the dark matter.

To better understand the layered structure of an aura, we can examine the formation process of a magnetic field. The magnetic field (aura) of every object consists of multiple layers, depending on the compression level of dark matter. The primary layer is composed of the dark matter directly enveloping the object's element cluster and contains the most intense energy fluctuations. The secondary layer consists of less compressed dark matter outside the primary layer, where energy fluctuations are less intense. The tertiary layer includes low-intensity fluctuations in the dark matter farther from the object. These layers constantly change based on the object's interactions with environmental factors, such as the magnetic fields of other objects or photon bombardment. For example, when a person is in a crowded environment, interacting with the auras of others, the energy fluctuations in their secondary and tertiary layers may increase, potentially causing sudden emotional changes like restlessness.

Some living beings and humans may react to the radiation they are exposed to. For instance, animals—especially cats—sensing a person's mood may result from their ability to detect these energy fluctuations. At the quantum level, the radiation emitted by a person interacts with the magnetic fields of other beings through the fluctuations it creates in the surrounding dark matter, leading to behavioral responses.

In conclusion, the aura is a scientific concept that emerged in history, whether intentionally or unintentionally. At its core, an aura is a magnetic field, and like magnetic fields, it consists of layers. The historical association of auras with colors may reflect the varying intensities of energy fluctuations in dark matter. The discovery of magnetic fields and dark matter might coincide with the first definition of the aura in history.

26. THE LAW OF INERTIA

The law of inertia, also known as Newton's first law of motion, explains one of the fundamental principles of motion. According to this law, an object at rest remains at rest, and an object in motion continues moving at a constant speed in a straight line unless acted upon by an external force. In other words, an object tends to maintain its velocity (both direction and magnitude). This principle can be summarized as follows:

- Tendency to Remain at Rest: An object that is not moving will remain at rest unless a force is applied to it.
- Tendency to Move at Constant Speed: An object in motion will continue moving at the same speed and in the same direction unless a force is applied to it.

The law of inertia can be observed in many everyday situations. For example, when a car suddenly brakes, the passengers inside lurch forward due to their inertia; they tend to continue moving at the car's previous speed before it stopped. Inertia is directly related to mass; objects with greater mass have greater inertia, meaning more force is required to change their state of motion.

In the Theory of Everything, the law of inertia does not exist at the quantum level because everything is constantly under the influence of a force. In other words, since everything is made of atoms and every space is filled with atoms, at the micro level, everything exerts a force on everything it contacts. Even in a multi-particle environment, inertia does not exist at the quantum level because every atom is in a state of vibration or positional change. According to the laws of thermodynamics, absolute zero—meaning complete stillness—is impossible.

At the macro level, an object that appears motionless is in that state because the positive force it exerts outward balances the inward positive force from dark matter. For objects moving at a constant speed, this motion is due to the momentum created by dark matter, including in the vacuum of space. (For a better understanding, you can revisit Hypothesis 7.)

Before explaining this further, I'd like to clarify that the phenomena described as inertia are not related to mass but rather to the strength of the magnetic field and the balance of forces between the object and dark matter. Consider a car moving at a constant speed with passengers inside. The object between dark matter is the car and its magnetic field. Every object inside the car, including the passengers, has its own magnetic field but is within the car's magnetic field. The magnetic fields of the passengers and the car are balanced relative to each other. However, because the car's wheels are turning, the pressure between the car and the dark matter outside it is unbalanced due to momentum. The car glides through the dark matter with its magnetic field cage. Everything appears normal until we disrupt this scenario by imagining the car crashing into a wall at 100 km/h. The moment the car crashes, it stops. As the car, which was gliding through the dark matter, comes to a halt, the dark matter accumulated behind the car exerts a forward positive force to return to its original position. This pressure violently pushes all the passengers forward, and this forward motion can cause harm to objects and living beings. For example, a passenger's arm might break because the elements in their arm cannot generate a magnetic field strong enough to resist the sudden pressure from the dark matter.

You can observe similar effects of dark matter when you strike a stationary object with force. You may have seen slow-motion videos on social media tagged with "inertia effect." Large objects create a strong magnetic field based on the structure of their constituent elements. Smaller objects inside or beside the large object are incorporated into its magnetic field and are protected. Therefore, when a stationary object experiences a violent impact from an external force, the large object with a stronger magnetic field reacts before the smaller objects within its field. This is due to the momentum effect caused by dark matter. For instance, if you quickly strike a glass on a table, the glass moves before the table does because the table's magnetic field is stronger than the glass's, and the momentum effect of dark matter causes the table to react more slowly.

Every object has a magnetic field based on the arrangement of its internal structure. When these magnetic fields approach another object, interactions and reactions begin. For example, even if there is a thin wooden board between a magnet and a metal object, the vectorial lines in the magnet's magnetic field exert a positive force, pulling the metal toward it. The elements in the wood and the dark matter within the wood are slightly affected by the magnet, resulting in a reaction imperceptible to humans. While the magnet appears to pull the metal without affecting the wood, the magnet's magnetic field does subtly influence the elements and dark matter in the wood. However, because the metal's magnetic field is stronger, the entire metal object is affected by this attraction. At the quantum level, this interaction may cause a slight increase in the vibrational frequency of the elements in the wood, though this increase is beyond human perception.

When two objects approach each other, their magnetic fields first interact, followed by the fusion of their magnetic field cage walls. For example, if a 1-ton flat-bottomed object is placed on a smooth surface, the magnetic field cage walls of the surface and the object fuse. When these two cage walls merge, the dark matter becomes highly compressed, binding them together. Moving the object in this state requires a significant amount of force. However, if you insert a lever under the object and apply a positive force, you loosen and separate the tight bonds between the magnetic field cage walls of the object and the surface. This allows you to move the object more easily with the lever. The ease of movement depends on the compression level of dark matter; for instance, if the elements in the surface were arranged more loosely, less force would be needed to move the object.

As seen, matter does not have a tendency; in nature, there are only interactions and reactions. At the quantum level, nothing possesses metaphysical emotions like desire or inclination. This demonstrates that the fundamental operation of nature involves constant force interactions.

27. RADIATION

In modern science, radiation refers to energy emitted in the form of waves or particles. It can originate from various natural and artificial sources and is divided into two main categories: ionizing and non-ionizing radiation. Both types have different applications, effects, and health risks.

Ionization is the process of converting an atom or molecule into an ion, which is why I will propose a hypothesis to redefine radiation. In modern science, the concept of ionization falls short of fully explaining the energy changes in atoms, as atoms are either in a solid, dense state with a small mass or in an expanded, softened state with a lighter mass relative to their size. Therefore, I will address radiation more comprehensively within the framework of the Theory of Everything.

In the Theory of Everything, radiation is defined as everything emitted outward from the cluster forming an object, affecting its magnetic field. This emission occurs at the quantum level through the vibrations of elements and soft atoms (electrons and photons). For example, when you burn a piece of wood, the light and heat from the fire are the electrons emitted by the wood, while the smoke consists of emitted elements. During this process, the wood's elements expand, transforming into soft atoms (electrons), which are then emitted outward from the wood's magnetic field. If you touch the burning wood or its flame with your hand, the magnetic field cage wall of the burning object damages your own magnetic field cage wall. This damage disrupts the molecular arrangement of the elements in your hand, causing harm, as the high-energy electrons emitted by the wood overcome the resistance of your magnetic field. As a result, by hearing the sound of the burning wood, smelling its smoke, seeing its light, and feeling its heat, you are exposed to all types of radiation. For instance, the sound emitted by the wood results from vibrations interacting with the dark matter in the air, which reach your ear's magnetic field and are perceived as sound.

The effects of radiation can be beneficial or harmful depending on its intensity. Everything you can perceive emits radiation that affects you. However, these effects remain harmless as long as they are limited to mild reactions at the quantum level. Radiation that cannot penetrate your magnetic field cage wall or the magnetic field of another protective object does not affect you. For example, the Sun emits intense radiation through photons, but Earth's magnetic field weakens this radiation. Earth's magnetic field acts as a barrier made of dark matter, scattering most of the Sun's photons. Nevertheless, some photons manage to penetrate this barrier and reach you, passing through your magnetic field cage wall and temporarily depositing external soft atoms into your body. These soft atoms interact with the elements in your skin, triggering chemical reactions that contribute to vitamin D production. However, if the Sun's photons penetrate the barrier too intensely (for example, due to prolonged unprotected exposure), your magnetic field cage wall's resistance weakens, and the elements in your skin can be damaged, leading to burns.

The impact of radiation is not limited to physical contact; environmental factors also play a role. For instance, a person near a nuclear power plant may be exposed to radiation from high-energy photons and elements emitted by the plant. If this radiation penetrates the person's magnetic field cage wall, it can disrupt the molecular arrangement of their body's elements, potentially causing serious health issues such as cellular mutations. However, if this radiation is blocked by a barrier with a strong magnetic field, like a lead shield, the person's magnetic field remains protected, and they are unharmed. At the quantum level, this process occurs because the elements in the lead shield, along with the dark matter, form a tight magnetic field that scatters the incoming photons.

Additionally, let me propose a mini-hypothesis: The internal quantum dynamics of humans cause them to continuously emit radiation, specifically photons. The element arrangement in every person generates vibrations at the quantum level, which are emitted outward as photons through their magnetic field. These photons accelerate due to the compression of dark matter by MRI machines (as you may recall, the speed of photons depends on the pressure exerted by dark matter). MRI machines create a strong magnetic field that compresses dark matter, increasing the speed of the photons emitted by the person. These accelerated photons are detected by the machine's sensors, enabling the device to generate images. For example, during a brain scan in an MRI machine, the vibrations of the elements in your brain may cause a more intense emission of photons, allowing the device to produce a clearer image. In this process, the compression of dark matter enhances the speed of the photons, optimizing the sensitivity of the detectors.

In summary, radiation encompasses everything emitted from physical entities, and its effects—whether beneficial or harmful—depend on its intensity. The impact of radiation is determined by quantum-level interactions and the resistance of magnetic fields, meaning environmental factors and protective barriers significantly influence its outcomes.

28. SPECTRUM AND WAVELENGTH

A spectrum is a sequence formed by the separation of light or other electromagnetic waves into their various components. It is a fundamental concept used across many scientific fields, such as physics, chemistry, and astronomy. As a powerful tool for understanding the structure and properties of elements and molecules, the spectrum has a wide range of applications in scientific research.

In the Theory of Everything, we continue to unravel the mysteries of photons. A photon is formed when solid atoms, rarely from dark matter but typically from elements, break off and expand. According to our theory, soft atoms are called electrons when they are close to an element, but they become photons when traveling through dark matter.

Since photons are composed of soft, expanded atoms, they are volumetrically light in mass and, due to their shapeless nature, exhibit superfluidity. A photon follows a path influenced by the pressure of dark matter. In a superposition state (for example, if the first photon passes to the left of dark matter, the second photon passes to its right), photons rapidly emitted from a source display wave-like behavior despite their particle nature. Because photons travel through the arrays of dark matter, they lose members along the way but can compensate for this loss by acquiring members detached from other photons.

During its journey, as long as a photon does not collide with an element amidst the dark matter, it continues by gradually losing and gaining members. However, as the distance increases, the gap between the photon and other photons traveling parallel to it widens, reducing its ability to regain lost members. This leads to both an elongation of its wavelength and a distortion of the visual information it carries.

When a photon collides with a gas element while traveling through dark matter, a violent collision occurs. This collision can split the photon cluster into two, three, or four parts, depending on the size of the element it encounters. Phenomena such as the formation of a rainbow can occur during this process. This is how a spectrum, or the light's array, is formed.

Photons have mass, a fact evidenced by numerous natural phenomena, such as the variation in light's wavelengths or the way photons (like lasers) wear down an object upon impact. Although a photon is relatively light compared to its volume, its effects are observable due to its acceleration speed.

29. DIMENSIONS

Everything we perceive in nature is physical at the moment of perception. However, once our perceptions of nature turn into informational memories, they become metaphysical. In other words, all information is metaphysical.

Our visual perceptions consist of three dimensions: length, width, and depth. All other dimensions, including the so-called space-time dimension, are metaphysical. Some might ask, "We observe changes in nature, so how can you not accept time as a dimension?" Time is the sum of the algorithmic processes our brain performs while we are awake. When we sleep or faint, the concept of time ceases to exist for that being. For this reason, I do not accept the changes in nature being defined as a "time dimension." Time, like all mathematics, is metaphysical. It is a unit of measurement, similar to meters or kilograms.

The concept of dimensions has no direct counterpart in nature. Everything we see is the result of reactions occurring in our brain. No, we are not living in a simulation; we exist in a physical nature. However, the algorithmic reactions in our brain would not have started without external stimuli. In reality, we live in a nature that has no absolute void—it is entirely filled. As I've explained since the beginning of the book, nature is covered with three main classes of indivisible atoms: dark matter, elements, and photons. If everything is filled with atoms, how do we see? We can see precisely because everything is filled with atoms. If there were an absolute void in nature, nothing could move, as there would be no energy to trigger movement. Despite everything being filled with atoms, the soft atoms formed slip through the dark matter, moving between clusters of solid atoms to reach our eyes. The photons entering our eyes transform into electrons, enabling the formation of images in our brain. For example, countless photons from the book you're holding right now bounce off in billions of pixels, entering your eyes and triggering reactions in your brain. So, we are not living in an illusion; everything we see has a counterpart in nature.

We cannot directly see dark matter because photons travel through dark matter. We also cannot directly see elements because elements are the source of light; the photon clusters reflected from elements pass through our eyes and create reactions in our brain. While we cannot directly see the photons carrying visual information, we can observe their massive forms. Photons are visible during neutron star formations, star and planet formations, as well as in lightning, thunderstorms, and electric sparks.

Returning to dimensions, the factors that create length, width, and depth in the objects we see are determined by the number of soft atoms forming the photon cluster. For instance, imagine looking at a red car. (You can also lift your head from the book and look at anything around you.) The car is not a photon source but an object that reflects photons. Photons from another source strike the car. If the light comes from behind us, each photon hitting the car shatters like a water droplet. Photons striking the car's width and length carry more soft atoms due to the angle of reflection, making the red color appear brighter and more vibrant. On the other hand, photons hitting the car's depth carry fewer soft atoms due to the reflection angle, altering the shade of red and making it appear less bright.

We perceive very small or distant objects as two-dimensional. This is because the photons striking the depth of such objects either cannot reach our eyes or carry so few soft atoms that they cannot be perceived as pixels.

30. CREDIT CARDS, RADAR, AND RADIO

Imagine a flat wall in front of you, with machines lined up behind you in a setup capable of firing thousands of baseballs at once, and a building ten times larger than the wall! Let the machines represent elements, the baseballs they fire represent photons, and you be a detecting device. When the machines fire the balls continuously, some of them bounce off the smaller wall and reach you. This is how radar works. A device sends a photon bombardment over a wide area. When some photons bounce back after hitting an object, the device detects them, confirming the presence of that object.

Now, imagine the wall in front of you is made of concrete pillars arranged at varying intervals and painted in different colors. Think of these pillars as a unique lock, like the barcode, card number, or QR code on your credit card, specific to the cardholder. You are the receiving terminal, and the baseballs are photons. When the baseball bombardment begins, only the balls that hit the pillars and get stained with different colored paint reach you. You can think of these paint stains as encrypted codes. Credit cards work in a similar way. The transmitting device in the terminal sends a photon bombardment at a specific frequency. Some photons bounce back from the barcode-like lock on your card. The receiving device in the terminal opens the cardholder's specific webpage or room on the server, withdrawing money from your account. For contact cards, the same system operates through physical contact. The terminal locates the cardholder's unique room; when you enter the PIN, the door opens, and the transaction is completed.

Radio operates a bit differently. Let's turn the ball-firing machines toward the sky. Fire the baseballs—representing photons—with enough force to reach space. As the baseballs travel through the atmosphere, they break apart due to friction, releasing their internal air into the vacuum of space, and their fragments fall back to Earth. Imagine the inner parts of these fragments contain something representing information. This is how radio signals work. Photons are fired toward space, but some, unable to penetrate Earth's magnetic field cage wall, reflect back (as I explained in the context of galaxy, planet, and star formations).

From the Earth's surface to the sky, a photon bombardment consisting of 100 million soft atoms each is launched. Due to intense interactions in Earth's atmosphere, photons pass through loosened dark matter—caused by the deformation of surface solid atoms—and collide with undeformed dark matter arrays in space, breaking apart. A photon composed of 100 million members splits into smaller photons of 10 million members each, scattering back toward the Earth's surface. A small fraction continues into space. This process is similar to diving into the sea from a great height. The photons, spreading over a wide area, are detected as radio signals, allowing you to listen to the radio. This creates multiple frequencies, and by decoding these frequencies, you can change the radio channel.

Attentive readers might ask, "If a photon formed with 100 million members in the Sun can enter Earth's magnetic field cage wall with 90 million members, why do most members of a photon starting with 100 million members from Earth fail to penetrate the magnetic field cage wall?" The answer lies in the pressure difference of dark matter. For example, transitioning from sea pressure to air pressure is easy, but moving from air pressure to sea pressure is difficult. The pressure of elements, photons, and deformed dark matter on Earth is significantly lower than the pressure of minimally deformed, solid dark matter in space. Contrary to common belief, the dark matter pressure in space is much higher. The highest pressure zones are found, in order, at the center of the universe, in black holes, and in dark energy regions. This pressure difference is the reason photons reflect back from the magnetic field cage wall, as I explained in the context of star and planet formation.

31. THE MECHANICS OF LIGHT PASSING THROUGH OBJECTS

Photons and electrons move or chart their paths based on the positive pressure they experience from dark matter. Photons and electrons can easily travel from a high dark matter pressure area to a low dark matter pressure area, but the reverse—moving from a low dark matter pressure area to a high dark matter pressure area—is difficult or sometimes impossible. This phenomenon is similar to the pressure difference experienced when entering and exiting the sea. The same principle applies to all objects composed of element clusters. The dark matter, elements, and electrons that make up these objects create countless combinations of pressure differences.

In this hypothesis, I will use crystals as an example. First, let me remind you that the smallest indivisible particle of everything is an atom, and everything is filled with atoms. For this reason, even when a light photon travels at the speed of light, an absolute void never forms. Crystals are molecularly bonded objects belonging to the three main classes I previously defined: photons/electrons, elements, and dark matter. When crystals form, they transition from a chaotic, vibrational, and dynamic state at the quantum level to an ordered state. For simplicity, we can think of this as a transition from hot to cold, similar to water turning from liquid to solid ice or lava solidifying. Some elements, when heated or cooled relative to their size, transform from a solid to a liquid state in a dynamic environment, eventually becoming crystals. During this transformation, the surface of some dark matter deforms—deformation of dark matter means the solid tips of the atoms on its surface become soft—and detaches from its array to become part of the cluster. In summary, crystals form when elements of a certain size trap deformed dark matter detached from its array.

This is why crystals have a weak magnetic field force and a dense electron structure.

When a light photon from an external source strikes the unique structure of a crystal, the mechanism I explained in the electricity section comes into play. Similar to the train analogy, most of the light photons striking from one direction pass through the crystal's weak magnetic field cage wall and join the cluster. The photon entering the cluster from outside transforms into an electron. These electrons push other electrons, expelling a number of electrons equal to those that entered, which then exit the other end as photons. The electrons expelled from the cluster transform back into photons, creating colors like those of a rainbow.

Contrary to popular belief, no photon passes directly through any object.

32. THE CAMERA OBSCURA PRINCIPLE AND THE CAMERA

The camera obscura principle relies on the linear propagation of light to function. This principle provides fundamental insights into light and image formation.

As I mentioned earlier in the Theory of Everything, photons do not travel in a linear or perfectly straight path; the structure of dark matter does not allow it. This is as I explained in the superposition section. A photon starts as an electron at the moment of its formation. It becomes a photon when it begins its journey through dark matter from its source element. As it moves away by pushing the electron near another element next to its source, it transitions into an electrical form relative to its exit from the cluster.

A laser beam is the only form where light appears to travel in a straight line. In a laser beam, the element produces a cluster of soft atoms in a single burst, with a volume close to that of dark matter. When the photon's volume is roughly the same as dark matter, it vibrates and travels in a straight path. However, when its volume is smaller than dark matter, it moves in a zigzag pattern; in other words, as it shrinks, its wavelength elongates.

Returning to the camera obscura principle, most light is the radiation of elements—that is, the solid atoms forming the element expand, detach from the element cluster, and scatter around. Everything we see is either radiation (light/photons) coming directly from an element to our eyes or photons formed from another source that bounce off an object. Photons, though singular in clusters, form rapidly and in countless numbers, making them superfluids that always create a wave function. During its journey, a photon loses members; in short distances, it compensates for this loss by incorporating members lost from other photons into its cluster.

However, as the distance increases, the number of members it can recover decreases. For example, at a distance of 10 km, if it loses 10 members, it might gain 9; at 20 km, if it loses 20 members, it might gain 17.

Imagine you're in a dark room on a sunny day, and you open a small hole in the wall. Instantly, a magnetic field cage wall forms at the hole due to dark matter. This magnetic field belongs to the outer wall of the room. Since the outside of the wall is bright and the inside of the room is dark, the magnetic field cage wall at the hole takes the shape of an invisible lens or mirror. As a result, all photons coming from outside invert as they pass through the hole and reflect onto the wall inside the room. The reason for this inversion is the higher pressure of the magnetic field cage wall formed at the hole. If we think of the first cameras instead of a camera obscura, the same mechanism would apply. Photons passing through the camera lens would invert, strike the plate inside the camera, and the impact would alter the position of the elements on the plate.

The mechanism I've described in this hypothesis has never changed. The same mechanics govern image formation in our eyes, modern cameras, and smartphones; only the internal algorithmic processes differ.

33. X-RAY MACHINES AND SIMILAR DEVICES

No soft atom passes directly through a living being or an object.

You might think I'm repeating the same things, but I believe this helps you better understand events in the quantum realm. It also shows how countless combinations arise from clusters formed by atoms with only two extreme states.

In X-ray machines, a photon bombardment is directed at a person. Since this bombardment is intense, the photons penetrate the person's magnetic field cage wall, join the cluster, and transform into electrons. As in the mechanics of electricity, these electrons begin to push each other. Because the photon bombardment continues steadily, the kinetic energy of the photons sustains a positive force. As each electron pushes the next, the electrons forced to the other side of the person exit the cluster and transform back into photons. However, solid structures like bones inside the person allow fewer electrons to pass through. As a result, the electrons at the bone level emit smaller photons as they exit. The detector captures these photons, enabling the X-ray image to be produced.

34. TELEVISION

You may have watched or are currently watching television. Technology continues to advance despite the many misconceptions that modern science is based on. Technological progress is largely achieved through trial and error. Theoretical physicists or philosophers propose theories, and millions of people attempt to disprove the trending ones. These efforts to disprove theories also drive technological development. For instance, even if all known theories were erased from human memory, technological progress would still continue. Most people, especially engineers, don't seem to care much about relativity or quantum theories. They focus on solving physical problems and ensuring the technologies they develop yield results. Nevertheless, theories remain important for everyone because they offer new perspectives, helping people devise innovative experimental methods.

In the Theory of Everything, everything that comes into contact forms a cluster. A cluster may contain other clusters or even empty clusters. When you plug in your television, you join the cluster of the electrical grid; when you use the internet, you join the cluster of the internet network. The moment you turn on the television, the electrons on the screen transform into photons and enter your eyes. Those electrons were once generated in hydroelectric, nuclear, wind, or solar power plants. Depending on the source of electricity for the grid you've joined, the electrons reach the television. Algorithmic processes occur within the television, resulting in the formation of images.

35. CELL PHONE SCREEN

Cell phones and computers are marvels of technology. The quantum mechanics of every hardware component that powers televisions and computers can be easily explained with the Theory of Everything.

Portable electronic devices like cell phones store electrons in a battery or a single cell. The electrons accumulated in the battery create images on the screen through algorithmic processes between hardware components. These electrons transform into photons on the screen, passing through your eyes and undergoing algorithmic processing in your brain. Additionally, signals received via Wi-Fi are detected by the phone. When these signals connect you to the internet network, the point where your finger touches the screen triggers algorithmic processes.

36. CINEMA

Image formation in cinema and similar systems differs from that in television. In digital projectors, a direct photon bombardment is sent to the screen. The photons reflected from the screen enter your eyes, creating the image. In analog film projectors, the device continuously emits photons at the same frequency. As the photons pass through the film, they transform into electrons. Once they become electrons, the mechanism I described for X-ray machines comes into play: the electrons transform back into photons on the other side of the film strip, depending on the element structure and arrangement of the film. The photons emitted from the device at the same frequency shift into different frequencies based on the position of the elements in each pixel. This allows you to enjoy the cinematic experience.

37. SENSORY PERCEPTIONS

In our social lives, we communicate with other people in countless ways. However, we are in constant communication with nature. You can escape from all people, but you cannot escape from nature; the moment you cut off all communication with nature, you are essentially dead. Most of our communication with nature occurs through photons. Of our five senses, sight and hearing interact with photons, while smell, taste, and touch interact with elements. We also have a sixth sense, through which we communicate with dark matter. For example, as I explained in the principle of inertia, when you're in a car and the car moves forward, the force pushing you back, or when the car suddenly stops, the force pushing you forward, is the pressure of dark matter. Sometimes, feeling that something will happen before it does, sensing that we're being watched, or noticing someone approaching could be due to the influence of dark matter. Humans' communication with dark matter may have diminished, but in some animals, this connection remains strong. For instance, migratory birds can sense the vector lines created by dark matter in the magnetic field. A lizard can cling to a flat surface. When you try to catch a fly, even if it's at an angle it cannot see, the fly can detect the ripples in dark matter and escape.

The influence of dark matter is everywhere, and every living being feels it. However, in some small creatures, this influence has become their most crucial sense for survival or hunting. While the effect of dark matter exists in everything, only humans on Earth can unravel its mysteries, transform them into knowledge, and share it with others. In this book, I have only tried to explain the Theory of Everything. Spreading this theory to the world is now the responsibility of influential people.

38. ACIDS

In the Theory of Everything, anything can be an acid to something else. I define an acid as a separator. If something is being deformed, melted, or burned, a separation process is occurring. For example, elements and photons in space act as acids to dark matter because they soften the tips of the solid atoms on the surface of dark matter. Elements and photons, which are asymmetrical relative to dark matter, deform dark matter as long as they remain in contact. But don't worry! Elements and photons—except in extreme cases like nuclear bombs, the Big Bang, and supernova phenomena—cannot break apart dark matter. Moreover, since space is vast and continuously expanding, elements and photons cannot destroy it. Humans are also exposed to acids, but in most cases, they remain unaffected.

To better understand acids, you should know the simple principle I've established: In clusters made of symmetrical members, an asymmetrical element acts as an acid. When a cluster with symmetrical members joins a cluster of asymmetrical members, the symmetrical cluster becomes the acid. This simple principle alone creates countless combinations in nature.

For example, life began with viruses. A virus is a cluster of asymmetrical elements rigidly bound together in a solid state. When a virus is exposed to a photon bombardment at the right frequency (a frequency composed of symmetrical photons), an acid effect begins through separation, and the virus transforms into a cell. In this context, anything can be an acid to something else.

Anything that disrupts the order of a stable system or shifts it from a static to a dynamic state is an acid. Anything that causes cancer or makes us sick is an acid.

Here's another example: Iron is a symmetrical cluster of large elements. Clusters made of lighter elements, like nitric acid, create an acid effect on iron. Additionally, asymmetrical elements that penetrate the magnetic field cage wall and join the cluster can also produce an acid effect. This is one of the reasons iron rusts when exposed to air.

Electrons and photons, which are clusters of soft atoms, are nature's most powerful acids. Photons, at the right frequency, can disrupt the structure of anything. However, in space, only soft atoms are immune to acid effects. This is because soft atoms—being shapeless, soft, light in mass, and superfluid—are the fundamental source of entropy. Due to geometric constraints, it is impossible for a soft, expanded atom to return to a solid state and shrink. The acid for a solid atom, on the other hand, is negative pressure generated by a positive force. When a solid atom is exposed to negative pressure, it expands and increases in volume, causing its structure to break down.

39. WEIGHT

What is weight? Why is something heavy? When you look into it, you'll typically come across this answer: Weight is the force exerted on an object due to gravitational pull. It is related to the object's mass and the gravitational acceleration of its location, measured in Newtons (N). In other words, weight is the force an object experiences under gravitational pull, and this force is directly proportional to the object's mass and the gravitational acceleration of its location.

This answer never satisfied me.

According to the Theory of Everything, weight is the ratio of force experienced by the element cluster in the losing direction during a battle of positive forces applied by dark matter on the element cluster from all directions. For example, if the dark matter closer to the ground wins this force battle, you would fly off into space. How would you then explain the weight of something that is floating using current definitions? It's explained in a rather amusing way, but let's not digress.

First, let's recall: All living beings and objects on Earth are clusters made of elements, electrons/photons, and dark matter. The gas layer forming Earth's atmosphere is also a cluster, and within our atmosphere, there are countless clusters. Clusters where dark matter is more dominant than elements are gases; clusters with roughly equal amounts of elements and dark matter are liquids; and clusters where elements are more dominant are solids. These form the three states of matter. If you cluster large elements without any dark matter between them, you get a nuclear atomic bomb; if you cluster them with lighter elements, you get a nuclear hydrogen bomb.

Clusters exert a positive force from the inside out, while the dark matter on the outside applies a positive force on the elements from the outside in. As the number of dark matter particles increases relative to elements, the cluster becomes lighter. A cluster dominated by dark matter can more easily resist the external inward force from dark matter. This is why, when you fill one bottle with water and another with gas, the water-filled bottle is heavier.

What is called the vacuum of space is filled with dark matter, but the dark matter pressure exerted on objects there is both higher and more balanced compared to Earth. This is why space is said to have weightlessness or microgravity.

40. GRAVITY

In the Theory of Everything, gravity is not a fundamental force—in fact, gravity doesn't even exist. However, since the concept of gravity is so ingrained in people's minds, I've occasionally had to use it metaphorically to make the subject more understandable.

According to the Theory of Everything, there is only one fundamental force: the positive force, which is the resistance exerted by an atom from the inside out. Since every atom possesses this force, everything is covered by a single fundamental force, driven by atoms exerting positive force outward. You might ask, "If there's only one fundamental force, how do so many secondary forces arise?" Every secondary, tertiary, and quaternary force in nature has an explanation. At the macro level, all forces are caused by force and shape asymmetry.

Let's draw an analogy. Imagine three people of the same weight and strength. These three people hold hands to form a circle and push each other with positive force, using the same level of strength. Assume these three individuals are symmetrical in both strength and structure. Atoms are like these three people—symmetrical in structure and strength, but in far greater numbers. Now, let two of these three people team up and apply positive force to the remaining individual. The lone person resists by applying a counterforce. Who wins? Naturally, the two who teamed up, right? The Big Bang was also a result of this symmetry breaking. During the Big Bang, an imbalance in the positive pressure exerted by dark matter on each other created negative pressure, leading to the formation of the first soft atom in space and initiating entropy.

Gravity does not exist in nature; it is a conceptual misinterpretation of an event. As I mentioned at the beginning of the book, the fall of an apple to the ground could just as well be interpreted as "sky thrust." The Theory of Everything explains the apple's fall through the dominant positive force. This is my interpretation. What matters is that the theory systematically explains everything. After this theory, anyone can continue to adopt whichever theory they prefer. Once the book is published, I'll feel at peace knowing I've fulfilled my duty.

So, why did the apple fall to the ground or move downward? Let's conduct a simple experiment, but I'll warn you: this experiment will later turn into a thought experiment. Find two sticks and an apple—or something of similar size. Imagine the apple as a cluster of elements and the sticks as representing dark matter. Use the ends of the sticks to apply positive force with your right and left hands, squeezing the apple. As you can see, the force applied by your right and left hands prevents the apple from falling to the ground. If you increase the force with your right hand, the apple moves to the left; if you increase the force with your left hand, it moves to the right. Now, imagine the apple being pushed by sticks from all directions. Whichever direction's positive force dominates, the apple will move in the opposite direction of that dominant force. This experiment was designed to help you understand positive forces and force balances. However, this alone doesn't fully explain why the apple moves downward. As I discussed in Hypothesis 39, one of the factors determining weight is the density of dark matter. In the apple, the dark matter density is lower compared to gases, so the dark matter in the upward direction wins the force battle, causing the apple to fall to the ground.

41. GRAVITATIONAL MASS

The concepts of gravitational mass and gravity are often confused, but they have distinct meanings. Here's how they differ: Gravitational mass is the attractive force exerted by all masses in the universe on each other. This force is directly proportional to the size of the masses and inversely proportional to the distance between them. Gravity, on the other hand, is the attractive force exerted by the Earth on objects at its surface, a force derived from the Earth's gravitational mass.

However, as I've repeatedly stated in the Theory of Everything, nothing that does not come into contact can attract anything else. Even when things do come into contact, without a fundamental positive force, nothing moves toward anything else.

The cause of macro-level events and phenomena in the universe, commonly attributed to gravitational mass, is actually a vortex. These vortices, as I explained in the section on planet formation, are dark matter vortices created by the massive photons forming the core during the formation of planets and stars. Each star system has its own unique vortex. Within the vortex created by the star, the planets in that star system also have their own vortices. The reason some planets rotate on their axes differently from others is the varying momentum of the massive photon core during its formation. This difference causes the dark matter vortex to rotate differently as well. The force that keeps stars within a galaxy is also the effect of this vortex. However, the vortices of black holes are different. In black holes, dark energy flows like water draining through a sinkhole, creating a vortex that engulfs all the star systems in the galaxy around it.

In the Theory of Everything, there is no gravitational mass at the macro level. The vortex factor in this theory prevents the formation of the three-body problem. At the micro level, the phenomenon of gravitational mass arises from geometric reactions that depend on the size and shape of elements.

42. GALILEO'S EXPERIMENT

Galileo Galilei's free fall experiment holds a significant place in the history of physics, demonstrating the fundamental principle that the speed of freely falling objects is independent of their mass. It is said that Galileo conducted this experiment by dropping objects of different masses from the Leaning Tower of Pisa, aiming to show that heavy and light objects fall at the same speed in a vacuum.

However, according to the Theory of Everything, the density of dark matter within the cluster of an object or living being is what determines its weight. A cluster moves through dark matter along with its magnetic field. But when the cluster accelerates, it does not move with its magnetic field; instead, it slides along with its magnetic field cage wall. The magnetic field cage wall is located right at the surface of the elements forming the cluster, while the magnetic field extends over a broader area. For example, the reason the Moon doesn't break away or lag behind during the Earth's orbital acceleration around the Sun is the Earth's magnetic field vortex. Caught within this magnetic field vortex, the Moon neither drifts away from the Earth nor approaches it to collide.

To analyze Galileo's experiment more effectively, a cannonball and a feather should not be dropped close to each other during free fall. Additionally, in a free fall experiment, a hollow cannonball with a core at its center could be spun on its axis and dropped from space to Earth alongside a light material. Would a vortex form in the magnetic field of the iron? This should be tested.

43. THE POWER OF THE WIND

Wind is the horizontal movement of air molecules in the atmosphere, typically resulting from their movement from high-pressure areas to low-pressure areas.

I have no objection to this definition, but I'd like to explain the mechanical process of wind according to the Theory of Everything. When dark matter in one region generates a positive force, a negative force forms in another region. Hot and cold are shaped by this law. In this context, the formation of high- and low-pressure areas supports this law. Earth's atmosphere is frequently influenced by the Moon's approach and retreat from Earth, its orbital position, sunlight from the Sun, and solar flares, though it is less often affected by the geometric alignment of planets. The dynamism in Earth's atmosphere largely stems from extraterrestrial factors. For example, the Moon follows a path around Earth: it approaches and moves away from Earth while moving up and down in the north-south direction relative to Earth's equatorial line. (Earth performs the same motion around the Sun.) This path of the Moon can trigger earthquakes, hurricanes, or volcanic eruptions—sometimes even all of them in a chain reaction. When the Moon approaches Earth, it applies a positive force, causing the dark matter between the Moon and Earth to compress. This compression of dark matter in space affects Earth's atmosphere, creating high-pressure areas in the dark matter. This is similar to pressing a water-filled balloon with your finger, causing another part to bulge. When a high-pressure area forms in one region, a low-pressure area emerges in another. The positive force exerted by the Moon as it approaches Earth also affects the dark matter in the atmosphere. Depending on the density of elements, this causes gaseous elements to condense into visible clouds. The cellular movement of clouds triggers further chain reactions. The Moon's path relative to Earth's equatorial line causes winds to veer to the right in the Northern Hemisphere and to the left in the Southern Hemisphere.

Earth's rotation on its axis causes day and night. The photon bombardment from the Sun penetrates the dark matter in the daytime regions, exerting a positive force from the inside out, which causes the atmosphere and Earth's surface to expand. The atmosphere, expanded by sunlight, experiences increased positive pressure from dark matter on Earth's nighttime side, making the night colder than the day.

Earth's up-and-down path as it orbits the Sun creates the seasons. This motion also plays a role in the formation of hot and cold temperatures. When it's summer in the Northern Hemisphere, the Earth's crust and atmosphere in that hemisphere expand in the east-west direction, resulting in longer daylight hours. At the same time, the opposite occurs in the Southern Hemisphere: it contracts, and the day-night cycle becomes colder, as I've described.

Many atmospheric events, including wind, as well as earthquakes and volcanic activity, are primarily driven by extraterrestrial factors. Light winds are the flow of elements in the air through dark matter. However, if extreme forces arise, elements can flow along with dark matter. When dark matter flows together with elements, the wind sweeps across the Earth's surface, capable of lifting objects weighing tons.

44. THE BUOYANT FORCE OF WATER

The buoyant force of water operates through a mechanism similar to that described in the hypotheses on gravity and weight. Water or liquids form when dark matter particles detach from their arrays and surround the entire structure, similar to ball bearings in a race. In the transformation of solid objects made of large elements into a liquid form, dense electron formation or intense external photon bombardment plays a key role. Therefore, when electron production or photon bombardment ceases, heat energy production stops, and the solid object in liquid form reverts to its original solid state.

Whether on Earth, in the vacuum of space, or in the atmosphere, the principle determining the direction of an object's movement is the battle between positive pressure forces exerted from the inside out and from the outside in. Weight, on the other hand, depends on the proportion of dark matter within the cluster of an object or living being relative to the elements. On Earth, heavier objects move toward the ground, while lighter ones move toward the sky. When you throw an iron cannonball into the sea, it sinks to the bottom because the dark matter in the iron cluster occupies less space relative to the elements compared to water. However, when you throw a piece of wood into the water, it doesn't sink, because the dark matter in the wood occupies more space relative to the elements. This is the reason behind the phenomenon known as the buoyant force of water.

You might ask, "How do ships made of metal, weighing tons, float on water?" The answer is this: Even though the ship is made of metal, the air on its surface or inside becomes part of the ship's cluster. Consequently, the dark matter in the ship's cluster occupies more space relative to the elements compared to water. As long as the proportion of dark matter in the ship's cluster doesn't decrease, the ship won't sink.

Due to the molecular structure of water, it is the most mobile medium on Earth and the closest to the pressure conditions of space. Conducting experiments in the sea that we can't perform in space can help us better understand space. For example, a variation of Galileo's free fall experiment could be conducted underwater (with the free fall occurring from inside the water toward the surface). By simulating the vortices found in star systems, we can study the Sun and our planets. Setting up a Karmatan experiment in a water pool could help us better understand how we can move so easily when everything is filled with atoms. In particular, photon behaviors can be observed underwater, as the object most similar to a photon in both space and on Earth is an air bubble underwater. An air bubble underwater exhibits behaviors similar to a photon because both the photon and the air bubble are shapeless and travel by determining their path based on the pressure forces they encounter. By setting up an experimental apparatus underwater, the superposition behaviors in the relationship between dark matter and electrons can also be observed.

45. WHY DID DINOSAURS GO EXTINCT?

The extinction of dinosaurs was likely caused by a combination of catastrophic events. An asteroid impact, massive volcanic activity, and climate changes are considered the primary reasons for this extinction. These events led to the collapse of ecosystems and significantly altered life on Earth. The extinction of dinosaurs remains a key research topic in modern biology and paleontology.

However, what I'm truly curious about is this: How did creatures approaching 30 meters in length and weighing tons form in their environment? By answering this question, we might better understand how dinosaurs went extinct.

In the Theory of Everything, I proceed without using mathematics. Instead, I rely on a single fundamental force to explain everything that comes to mind. In spherical structures, the pressure from the sides decreases as you move outward from the center of the sphere. This is why we struggle to reach the deepest parts of the oceans or the Earth's core. The same principle applies to the atmosphere. Our ability to withstand this pressure is due to the elements, electrons, and dark matter within us exerting an outward positive force to resist the external pressure. When standing at sea level, positive pressure is applied from all directions. However, the pressure parallel to the ground is stronger relative to the pressure from above. This allows us to stand upright and enables humans to reach heights exceeding 2 meters. If the opposite were true, everyone would be a reptile or remain as short as 50 cm. As we ascend from sea level, the pressure from above begins to dominate over the pressure from the sides. In other words, the higher we go, the more space dark matter occupies relative to elements. Beyond the atmosphere, we enter a region entirely composed of dark matter. In this region, dark matter envelops the Earth like a cage. This cage wall doesn't easily allow light or element particles to escape. Let's assume the magnetic field cage wall begins where the atmosphere ends, at about 100 km altitude. For the atmosphere during the time that allowed dinosaurs to form and survive, let's assume a height of 70 km.

During the dinosaur era, the atmospheric layer was thinner than it is now. At that time, with a thinner atmosphere, the Earth's diameter was also slightly smaller. This structure indicates a high pressure parallel to the ground. The pressure from the sides being higher relative to the pressure from above made it possible for animals and trees to grow taller and larger. To put it in simpler terms, "Since gravity was weaker, gigantic dinosaurs and trees, like those in the movie Avatar, could exist."

But everything changed when a massive asteroid struck the Earth. This event triggered volcanic activity and a series of catastrophes. Millions of tons of elements from the asteroid were added to the Earth. The atmosphere, which was at 70 km altitude, rose above 100 km. When the dust clouds that covered the atmosphere settled on the surface, the Earth's current diameter was formed. Even the creatures that survived the initial impact couldn't stand up due to their massive size, as the pressure from the sides decreased, and they died. This led to a mass extinction. Millions of years later, a sudden explosion of life occurred, and most of the species that formed then, including us, continue to survive today.

In conclusion, the end of the dinosaur era and the reason dinosaurs never reemerged can be attributed to the strengthening of gravity. However, I don't like using the concept of gravity. According to the Theory of Everything, the elements from the asteroid that were added to the Earth's cluster increased its mass, reducing the pressure parallel to the ground and thus ending the dinosaur era.

46. DIRECT CURRENT

Direct current (DC) is a type of electric current where electrical charges flow in a constant direction. Batteries, solar panels, and some generators produce DC current. Many applications, such as electronic devices, electric vehicles, and batteries, use direct current. Unlike alternating current (AC), DC provides a stable voltage and is ideal for short-distance energy transmission. It is preferred in electronic circuits because it offers a reliable and consistent energy source.

In the Theory of Everything, everything in nature is composed of indivisible atoms. The solid, raw form of an atom is the smallest particle found in nature. The largest cluster formed by solid atoms is the symmetrical dark matter particle. When a dark matter particle breaks apart, elements emerge in asymmetrical shapes. Elements are also asymmetrical clusters of solid atoms. When a solid atom expands and increases in volume, it transitions into the soft atom form. Soft atoms, in turn, form numerically asymmetrical clusters.

If you notice, elements are asymmetrical in shape, while photons are numerically asymmetrical. This is because photons are shapeless and take on the shape of their environment. Whenever soft atoms are near any element, I refer to these soft atoms as "electrons."

Explaining direct current using an air-balloon analogy can make the phenomenon more understandable. Imagine an uninflated balloon as an empty battery and the air you use to inflate the balloon as electrons. When you inflate the balloon, it represents a charged battery. Pinch the balloon's opening lightly to prevent air from escaping. Now, press the inflated part of the balloon with your finger using a constant force. The air inside the balloon will exit through the opening at a constant speed (+) due to the positive (+) force you apply. The operation of direct current is that simple. However, batteries don't shrink in volume when their electron emission stops.

A battery is like a prison where electrons are compressed from the outside, similar to blowing air into a balloon. What prevents the electrons from escaping the battery is its magnetic field cage wall. When the positive and negative terminals of the battery come into contact with a conductive material, the cage wall holding the electrons temporarily deforms or opens. When you release the opening of an inflated balloon, the stretched balloon contracts, pushing the air out. In a battery, the expanded soft atoms—electrons—are compressed like a spring, so they are expelled from the battery as electrical energy.

I'm not an expert, but I can offer this advice to engineers working on electrical storage: Not every material can store electricity. At the quantum level, elements and dark matter must be in a loose and relaxed environment to allow externally introduced electrons to be compressed between them. The dark matter in such an environment must have a deformed surface. For example, the dark matter in glass and crystal formations has a deformed surface, which is why they easily allow light to pass through. Additionally, suitable elements must be used in battery storage. You engineers can figure out the details—that's not my area of expertise.

47. ALTERNATING CURRENT

Alternating current (AC) is a type of electric current where the direction of electrical charges changes periodically. This change typically occurs 50 or 60 times per second (Hertz) and forms the foundation of electrical grids worldwide. One of the main reasons AC is widely preferred is its ability to efficiently transmit energy over long distances. Transformers allow voltage levels to be easily adjusted, reducing energy loss and making distribution more cost-effective. AC is used to power most electrical devices in homes, workplaces, and industries, with applications ranging from lighting systems to large industrial machinery. Additionally, many critical components, such as electric motors and generators, are designed to operate with AC. The flexibility and efficiency of alternating current have made the widespread use of electricity possible, supporting technological progress.

In the Theory of Everything, everything is simple. Since I developed this theory, I'm quite familiar with its principles. For example, after watching and learning from a 5-minute video titled "How do direct current generators produce electricity?" for the first time in my life, I was able to unravel the quantum mechanics of the phenomenon in seconds. Once you master this theory, you too will be able to solve any mystery of nature in seconds.

The method of generating electricity for alternating current differs from that of solar panels. While direct current (DC) generators produce electrons directly from elements, solar panels generate electrons directly from light. Here's how an alternating current generator works:

Working Principle:

- Formation of the Magnetic Field: The stator of the generator creates a stable magnetic field, which can be provided by permanent magnets or electromagnets.

- Rotor Rotation: Mechanical energy (e.g., supplied by a motor or turbine) rotates the rotor. The conductive coils on the rotor move within the magnetic field.
- EMF Induction: As the rotor rotates within the magnetic field, an electromotive force (EMF) is induced in the conductive coils, according to Faraday's law. This EMF generates a current in the rotor coils.
- Alternating Current Generation: The continuous rotation of the rotor causes the direction of the current to change periodically, producing alternating current (AC). (In DC generators, a commutator converts this into a steady current, but in AC generators, the current remains alternating.)
- Output Current: The generated alternating current is transmitted to the external circuit via brushes.

The rotor's rotation within the magnetic field of a permanent magnet triggers photon production. This process is quantum-mechanically identical to the formation of lightning and thunderstorms. The dark matter in the rotor's magnetic field is looser compared to the dark matter in the magnet's magnetic field. As the rotor spins, negative pressure zones form between the loose and tighter arrays of dark matter. A micro-scale version of the lightning formation I described occurs here: electrons are taken from the elements in the air. These electrons transform into photons, join the conductive cable cluster in the rotor, and turn back into electrons. To put it more clearly: Just like naturally occurring lightning, photons—too small to be seen with the naked eye—form between the rotor and stator. These photons travel like lightning, striking the copper conductor, joining its cluster, and transforming into electrons.

The generated electrons are pushed into the grid, similar to the analogy of a water hose. However, the distance from the production site to the destination is very long. This long-distance challenge is overcome by transformers placed at regular intervals, which create negative pressure on the electrons (you can think of transformers as water reservoirs). Imagine a long hose with faucets or valves placed at intervals along its length. The electrons pushed between the elements forming the cable struggle to flow over long distances. Transformers create negative pressure by allowing the electrons to flow toward them. The accelerated electrons transfer their gained kinetic energy to other electrons by passing to the other side of the cable.

I don't believe this type of electricity generation would work in space. Even if magnetic fields rotate, there are no elements among the dark matter in space to facilitate electricity production. For now, solar energy is the most suitable method in space. However, if my theory gains traction, new methods of electricity generation might be invented. Experiments to test this should be conducted and observed in the vacuum of space.

48. BOSE-EINSTEIN CONDENSATE

The Bose-Einstein condensate (BEC) is a state of matter where, under specific conditions, atoms at extremely low temperatures combine to form a single quantum state. First theoretically predicted by Satyendra Nath Bose and Albert Einstein, this state was experimentally observed in 1995 by Eric Cornell and Carl Wieman. BEC is a phase in which atoms move collectively, exhibiting extraordinary properties like superfluidity and enabling the study of quantum mechanics at a microscopic scale.

However, in the Theory of Everything, matter exists only in three states: solid, liquid, and gas. Atoms, on the other hand, have two states: solid and soft. In mainstream science, many concepts have been manipulated either out of ignorance or to fit established theories. The phenomenon known as Bose-Einstein condensation is one of them.

First, at the quantum level, there is no such thing as hot or cold; hot and cold are physical perceptions of macro-level events. Heat is the perception of events resulting from the loosening of elements' positions when arrays of dark matter create a negative pressure (loose) environment between them. Cold is the perception of events resulting from the tightening of elements' positions when arrays of dark matter create a positive pressure environment between them. You can understand this more easily with the hand-and-fly analogy: Your hand represents dark matter, and the fly represents an element. When you close your hand, the fly's space narrows, and you can't hear its buzzing; the element doesn't emit electrons, and since there's no electron emission, you perceive cold. When you open your hand, the fly's space expands, and you hear its buzzing; the element emits electrons, releasing heat and light, and you perceive warmth.

In the Theory of Everything, absolute zero means that atoms and all clusters of atoms are completely motionless. However, this is impossible since everything is filled with atoms. Even before the Big Bang, when space contained only symmetrical dark matter, each atom was still in motion, resisting external positive force pressure by exerting an internal positive force.

Regarding the phenomenon called Bose-Einstein condensation: When arrays of symmetrical dark matter strongly suppress the dark matter in their own array and in other arrays, the intense positive pressure from the dark matter expels some elements from the liquid cluster—like squeezing water out of a wet sponge. The expelled elements begin to move based on their asymmetrical shapes and sizes. This is similar to the stick analogy I described in the gravity hypothesis. The elements that separate from the liquid and rise reach slightly looser dark matter arrays, turning into a gas. Then, they condense a bit more, transitioning into a liquid droplet form. In the experiment, the rubidium atoms (which are actually elements) don't freeze like water because they have symmetrical proportions relative to each other. Normal water, being composed of countless asymmetrical elements, doesn't exhibit the same phenomenon.

49. PLASMA

Plasma is defined as one of the four fundamental states of matter, consisting of ionized gases. At sufficiently high temperatures or under a strong electric field, atoms in the gaseous state lose their electrons, forming positive ions and free electrons. Plasma is found in stars, which make up most of the universe, in the Sun, and in natural phenomena like lightning. In industrial applications, it is used in technologies such as fluorescent lamps and plasma displays.

However, in the Theory of Everything, matter has only three fundamental states: solid, liquid, and gas. Plasma is, in fact, a gaseous state of matter. The notion that plasma is a fourth fundamental state stems from a manipulative interpretation aimed at fitting it into the frameworks of existing theories about nature's workings.

The phenomenon of plasma occurs in a highly negative pressure (loose dark matter) environment. An event similar to the loose hand in the hand-and-fly analogy I mentioned earlier is enacted here by dark matter. In a negative pressure environment, an element loses members as its solid atoms transform into soft atoms and escape as photons; these photons are observed as plasma. Plasma can form anywhere there is a sufficiently strong negative pressure environment and the right elements are present. I suspect the definition of plasma was based on the wavelength of light. Negative pressure environments are created by positive pressure environments. The area where plasma forms resembles an induction field.

50. LASER

A laser beam emits high-intensity, coherent light at a specific wavelength. The term "laser" is an acronym for "Light Amplification by Stimulated Emission of Radiation," and it produces light through stimulated emission. Lasers emit monochromatic and phase-coherent light, allowing them to create a very narrow beam. This property makes them widely used in communication, medicine, industry, and military applications. They enable data transmission in fiber optic communication, provide precise cutting and treatment in medical surgeries, and are used in industry for cutting, welding, and measurement tasks. Laser technology underpins many innovations, such as holography and laser printers. With advancing technology, the applications and impact of lasers are rapidly expanding.

In the Theory of Everything, the laser light photon may hold significant mathematical importance. I don't know mathematics, so I'll offer some suggestions to mathematicians, perhaps overstepping my bounds.

Photons are always clusters of soft atom particles. However, because photons are produced very rapidly at their source and immediately travel through dark matter, they create a wave function. Being shapeless like water, photons are fluid. Their relatively light structure compared to their volume makes them the superfluid of space. Photons are numerically asymmetrical: the number of members in a photon cluster decreases as the wavelength increases and rises as the wavelength shortens, creating an asymmetrical structure. In this context, we should classify photons based on the number of elements in their cluster.

Due to their lightweight structure relative to their volume and the shape of dark matter, which is a cluster of solid atoms, a photon particle can never follow a straight path. Even a laser photon, which appears to travel straight, moves through dark matter like a snake, vibrating as it goes. A light photon never travels in a spherical shape; it gains acceleration due to external pressure and takes on a snake-like form.

This is why the laser light photon is significant. Because its wavelength is short, a laser light photon, when it first detaches from an element, may be the same size in volume as a dark matter particle. If physicists use the laser light photon as a reference point, they can calculate the number of soft atoms in each light wave. Using this method, they can determine the volume of a dark matter particle, which is a cluster of solid atoms. Through deductive reasoning, they could mathematically solve many mysteries, such as the mass and diameter of a solid atom, the number of atoms in elements, and much more.

Suggestion: The laser light photon can be used as a reference point to classify elements and photons, determine the volumetric size of dark matter, and calculate the size of an atom.

51. WATER ON EARTH

The origin of water on Earth is a debated and researched topic in the scientific community. Two main hypotheses attempt to answer this question:

- Release of Volcanic Gases: In Earth's early history, volcanic activity was intense. During this process, water vapor and other gases were released into the atmosphere along with magma and other volcanic materials. Over time, this water vapor condensed, fell to the surface as rain, and formed the oceans.
- External Sources: This hypothesis suggests that water was brought to Earth by external celestial bodies such as comets, asteroids, and meteorites. These bodies collided with Earth during the early stages of the Solar System, carrying water with them. Carbonaceous chondrites, a type of meteorite, are known for their high water content and may have significantly contributed to Earth's water reserves through these collisions.

These two theories are the most widely accepted explanations for the origin of water on Earth. Scientific research continues to better understand their validity.

However, in the Theory of Everything, water did not come to Earth from elsewhere; all the water on Earth formed through the transformation of viruses. Viruses are formed by the bonding of asymmetrical elements in a solid state. Elements are more likely to bond with other elements that have a similar volumetric mass ratio, forming clusters or molecular bonds. They can also form molecular bonds with much larger elements, but these bonds can easily break due to factors like friction between larger elements—similar to the dust produced when rubbing two stones together or the rusting of iron. The clustering of elements through bonding is a widespread phenomenon in nature. However, this type of clustering differs from the formation of stars, planets, and moons. Stars, planets, and moons form when elements within the magnetic field created by a massive photon at the core are pushed toward the core by the positive force of dark matter. Virus formation, on the other hand, results from the direct bonding of elements that come into contact. In this context, the particles shed from our skin are viruses. Celestial bodies like comets and meteorites, which roam space without being bound to a planet, are also virus clusters.

A photon can act as an acid for everything. With the right frequency, photon bombardment can break down a virus, dissolving it like an acid. The water and life on Earth did not come from somewhere else where countless viruses existed; they formed through the breakdown of viruses that were already present.

52. HOW DID HUMAN RACES FORM?

Human races refer to populations living in different geographical regions of the world, distinguished by certain physical and cultural characteristics. These differences are evident in genetic traits such as skin color, hair type, and facial structure. Historically, people were often categorized into the following major groups based on physical characteristics:

- Caucasoid (White)
- Mongoloid (Asian)
- Negroid (African)
- Australoid (Aboriginal and Papuan)

However, today, these classifications have largely been abandoned due to scientific inaccuracies and social sensitivities.

In this hypothesis, I'll explain the formation of human races according to the Theory of Everything. Contrary to common belief, humans do not share a common ancestor. During the dinosaur era, the diversity of life ended in a chaotic environment caused by the convergence of various factors I previously described. Some time later, conditions became suitable for a new explosion of life. Countless species began to form instantly, depending on the local dark matter pressure force, the number of viruses, photon frequency, and the variety of elements in their environment. All living beings that reached adulthood, including humans, did not form through the division of a single cell but through a single cell assimilating external elements. These external elements, based on their symmetry and the pressure they were subjected to, were reactively sorted within the assimilating cell, forming new clusters in appropriate locations and thus creating organs. As organs formed, genders also emerged based on the ratio of elements.

During this explosion of life, human races formed simultaneously in their known geographical regions, shaped by the conditions of that time. The transition from a single cell to adulthood was very rapid. For example, while a human takes 9 months to reach baby size in the womb, in a natural, open environment, a baby size was reached in just a few weeks. This is because a single cell forming in the womb and one forming in the external environment are exposed to different pressures. In the womb, element assimilation is limited and slow, whereas in the external environment, element assimilation is much freer and faster.

Human races formed simultaneously, influenced by the environmental factors of their respective geographical regions. Even now, just like us and other living beings, single cells with the potential to become humans or other species are forming, but they are immediately hunted and consumed by adult organisms.

Suppose all living beings on Earth died simultaneously today. Not long after—perhaps tomorrow—a new explosion of life would occur, and countless new species would form again, starting from a single cell.

53. THE CONCEPT OF REPRODUCTION BY DIVISION

In this hypothesis, I'll address the concept of cell reproduction by division. According to the Theory of Everything, reproduction by division in nature is purely numerical. When a cell divides, two cells are formed; if both cells divide simultaneously, four cells are formed, and the process continues in this manner. Think of a water droplet as a cell: Gas condenses and forms a water droplet. When this droplet falls into a lake, it becomes part of the lake and loses its cell-like form. When you take a droplet from the lake, it regains its cell-like form. This is why you see cells when you take a drop of blood from a human.

Viruses are the fundamental building blocks of life. Each virus has a genetic structure or code made up of elements. When viruses begin to break down, solid atoms on the surface of solid elements expand—like bubble formation—transforming into soft atoms and becoming electrons. During this process, elements detach some dark matter from their arrays and surround themselves with it, forming a single cell. The moment a single cell forms, the resistance it exerts from the inside out compresses the symmetrical dark matter on its surface, automatically creating the cell's magnetic field and magnetic field cage. As the cell takes in external material, its volume increases, and its resistance grows, but the positive pressure from dark matter pressing inward continues. If the genetic structure of the expanding cell contains a high number of electrons, the cell cannot grow further and divides. This is because electrons, being soft, have weak resistance from the inside out. Due to the high concentration of electrons in the cell's structure, it can never become a plant, human, or another type of organism; it remains a bacterium temporarily, depending on environmental factors. Unless its genetic structure changes, it won't grow, and unless it reverts to a virus, it will remain a single cell.

54. NATURAL SELECTION

According to the Theory of Everything, natural selection begins in the single-cell stage for all terrestrial living beings, including humans, plants, insects, and trees.

We refer to the molecular arrays formed by small elements sticking to each other's flat surfaces in a solid state as viruses. Arrays formed by larger elements, on the other hand, encompass all other solid materials such as rocks, iron, gold, and crystals. When viruses are exposed to appropriate photon bombardment, they begin to break down (dissolve) due to positional changes caused by electron emission. During these positional shifts, if the elements detach some dark matter from their arrays and surround themselves with it, they transform into a liquid cell; if they cannot, they become a solid seed. In this process, natural selection also occurs.

How Does Natural Selection Occur?

First, let me briefly explain how male and female genders are determined. In mainstream science, gender formation is described as follows: "In many species, gender determination is typically controlled by genetic instructions. In humans and other mammals, male and female genders are generally determined by the X and Y chromosomes. Male individuals carry XY chromosomes, while female individuals have XX chromosomes. The genetic material from the mother and father determines the gender of the zygote." However, in the Theory of Everything, the gender of a single cell is determined the moment it transforms from a virus into a cell. If the X element is more dominant than the Y element, male organs form; if the Y element is more dominant than the X element, female organs begin to form as soon as the single cell starts growing.

A similar mechanism applies to the formation of all species in a given region. However, the mass and volumetric structure of elements in their genetic makeup differ. The key factor determining the species is the range of volumetric and mass sizes of asymmetrical elements. For example, let's metaphorically assume that a cell composed of elements in the 0-10 millimeter range grows into a human. If elements in the 10-20 millimeter range are introduced to this cell, it will not become a human but will instead transform into a seed. The most critical stage—before the organs form and the cell's magnetic barrier strengthens—is the range of element sizes the cell is exposed to while still in its cellular form. Natural selection occurs at this critical moment.

Let me provide another example: Suppose a cell is composed of elements in the 0-4 millimeter range. As this cell grows into adulthood, it becomes a bird with a sharp, long beak. However, if elements in the 4-6 millimeter range are introduced to the cell at this critical moment, the liquid cell turns into a solid seed. As the seed grows into adulthood, it becomes a plant that blooms flowers with nectar in its depths, which the bird favors.

Conclusion: Natural selection is shaped by the range of elements a single cell is exposed to during its cellular form. Once the cell or seed forms its main organs and grows a bit, it can never transform into another species. Contrary to popular belief, a fish does not evolve into a crocodile, a crocodile into a monkey, or a mouse into a lion.

55. FERTILIZATION OF THE EGG

The female pregnancy process typically begins when an egg travels through the fallopian tubes and reaches the uterus. Each month, during ovulation, one of the ovaries releases an egg cell. If this egg is fertilized by sperm during sexual intercourse, the fertilized egg reaches the uterus, implants itself, and pregnancy begins. Inside the uterus, the fertilized egg develops into an embryo and later a fetus.

According to the Theory of Everything, the formation of an egg occurs through the clustering of elements with a genetic code, filtered in the internal organs, under the body's internal pressure.

This hypothesis will be quite short. As I previously stated in the Theory of Everything, "Everything can be an acid to everything." Male sperm acts as an acid to the female egg. The elements forming the sperm are small and surrounded by a high concentration of electrons. This structure of the sperm creates an acid-like effect on the female egg, which is formed by the compression of solid elements under high pressure. This effect causes the egg to break down, initiating the pregnancy process. A single sperm doesn't swim and penetrate the wall; rather, countless sperm penetrate the egg's wall to start the process.

56. THE FORMATION OF LIFE IN WATER

We previously noted that the water on Earth formed from pre-existing viruses. In this hypothesis, I'll use the term "water" to refer to all bodies of water, including lakes, rivers, seas, and oceans. When I mention "water," understand that I mean both smaller bodies like lakes and rivers, as well as larger ones like seas and oceans.

According to the Theory of Everything, the formation of life in water differs from that on land because the pressure within water is significantly higher than on land. The formation of cells in water is quite different from that on land. On land, viruses transition into a liquid form to create cells, whereas in water, elements bond together to trigger the formation of solid and semi-solid organisms.

Water, as a liquid state of matter, consists of dark matter particles that have detached from their arrays and are surrounded by light elements. These detached dark matter particles increase their resistance by exerting a positive force from the inside out, countering the symmetrical dark matter pressure from the outside in. In doing so, they also exert pressure on other dark matter particles surrounded by elements in the water. This process is what makes the liquid fluid. Recall the ball-bearing analogy we discussed earlier!

Considering the vast area covered by water on Earth, it becomes clear that there is immense pressure on the elements within it. When external factors—such as melting glaciers, dust clouds rising from land, and elements brought by rivers—are introduced into this high-pressure environment, some solid elements in the water recombine to form viruses. These molecules, which grow in volume as they turn into viruses, rapidly bond with other elements in the water, triggering the formation of microscopic marine life. Just as on land—indeed, many times more than on land—life in water continuously forms.

Countless organisms like algae and krill continuously form in most bodies of water without interruption. Suppose we removed and consumed all the living organisms in the water at once; the very next day, an explosion of life would begin in all the major bodies of water on Earth.

In Summary: The formation of life in water begins with viruses bonding through molecular connections. While viruses break apart to form cells on land, in water, viruses combine to form a single cell that begins to grow. The primary factor driving this is the high pressure created by water. The lack of structured arrays of dark matter in the water cluster continuously causes light elements to bond with each other. This bonding, in high-pressure environments, is what makes seawater salty.

57. THE SINGLE FUNDAMENTAL FORCE

Discovering the single fundamental force of nature would be a groundbreaking achievement in the scientific world. This discovery would reveal that all forces and interactions in the universe are different manifestations of a single fundamental force. The electromagnetic force, weak and strong nuclear forces, and gravitational force are merely different expressions of this fundamental force. The Theory of Everything offers a completely new perspective on understanding the universe's workings and helps solve unresolved mysteries across all fields of science.

According to the Theory of Everything, the single fundamental force of nature is resistance. I initially tried to announce my theory on social media under the name "Resistant Matter Theory." Resistance is a force we frequently encounter in our daily lives; it's a force we witness even while reading this book right now. It's astonishing that something so simple and visible hasn't been declared the single fundamental force. The reason for this lies in the scientific community's acceptance of disparate, unintegrated theories as the foundation of science. When a flawed theory is taken as the basis, finding the truth becomes difficult; each scientific advancement built on a flawed theory simply piles error upon error. Even if a theory explains millions of natural phenomena, if it fails to explain a single event or contradicts its foundational reference point, that theory should be discarded.

In the Theory of Everything, the single fundamental force is the positive force that an atom exerts from the inside out. We can also interpret this force as matter's resistance to annihilation, its opposition to external pressure.

You might ask, "Can a single fundamental force give rise to such complex natural phenomena and elements?" The answer is: Yes, it can.

How?

When an atom exerts a positive force from the inside out, it will expand the moment it finds space to do so. When the atom expands, it transitions into a soft atom form. Being soft and relatively larger in volume compared to another atom makes the environment more dynamic. As the saying goes, "One hand has nothing, but two hands make a sound." The force exerted by two atoms from the inside out overcomes the resistance of a single atom, causing its position to change. As the atom's position shifts, it pushes another atom it contacts while pulling yet another atom with negative pressure (due to adhesion). Thus, the single fundamental force—the positive force from the inside out—gives rise to a secondary force: negative force. When this single fundamental force combines with the secondary negative force it creates, it's no surprise that it accounts for the four forces misidentified in mainstream science: the electromagnetic force, weak and strong nuclear forces, and gravitational force.

This 57th hypothesis alone should revolutionize science, but in a rigid scientific community, this seems unlikely. Set me aside! What would happen if a world-renowned physicist came forward and declared, "The positive force exerted by matter from the inside out is the single fundamental force"? I wouldn't know. If I've managed to explain everything with a single fundamental force and have conveyed what that force is by this page of the book, then I'm content.

58. THE SINGLE FUNDAMENTAL ENERGY

According to the Theory of Everything, since the single fundamental force in nature is the positive force exerted by an atom from the inside out, the single fundamental energy in nature is the reactions resulting from the interactions of atoms.

Energy is the totality of interactions that arise from the movement of matter or atoms. If an atom does not move at all—which, as I explained earlier, is impossible because absolute zero cannot be reached—there is no energy.

I intended to elaborate on this hypothesis and connect it to an existing type of energy. However, from what I can see, all energy types defined by humans are structured as secondary, tertiary, or further derivatives of my definition.

59. ENERGY AND TIME

If you ask, "What is energy?" you might get answers like these:

- "The power inherent in matter, manifesting as heat or light; energy."
- "Energy is a fundamental concept that expresses the capacity to do work in physical systems."
  You might hear other responses, including some who say, "Energy is matter, and matter is energy."

According to the Theory of Everything, the concepts of energy and time are similar. As I previously stated, energy is the totality of interactions that arise from the movement of matter or atoms. Time, on the other hand, emerges when we are conscious, through the algorithmic movements of matter in our brain at the quantum level. When these algorithmic processes in our brain are intense, time seems to flow quickly; conversely, when these processes slow down, time appears to pass more slowly.

Our brain can perceive energy and understand time only when we are conscious. When we are asleep or in a coma, since we are unconscious—like a rock—we cannot comprehend either energy or time.

60. WHAT IS THE FUNDAMENTAL CAUSE OF ENTROPY?

Entropy is a measure of disorder or randomness in a system. It is associated with the second law of thermodynamics, which dictates the direction of natural processes: In a closed system, entropy will increase over time. Entropy reflects how widespread and unusable energy distribution becomes. As a system's entropy increases, energy becomes more disordered and dispersed. This concept plays a significant role in physics, chemistry, information theory, and other disciplines. Entropy helps us understand the balance between order and chaos in the universe, offering deep insights into the fundamental workings of nature. However, these meanings and definitions often serve as a refuge for many physicists when faced with difficult questions. If you ask a physicist, "What is the fundamental cause of entropy?" or "Why is entropy irreversible?" you're unlikely to get a clear answer. Let's explore the quantum-level source and cause of entropy according to the Theory of Everything.

Entropy began with the changes that occurred in infinity before space was formed, and it will continue eternally. In our nature—that is, in space—the fundamental source and cause of entropy is the process by which an atom expands from a solid state into a soft state. Soft atoms cluster as electrons and photons, behaving in accordance with the pressure they are subjected to.

Once an atom expands and increases the volume it occupies, it is impossible for it to compress and return to its original solid state. This cannot be achieved using symmetrical and asymmetrical solid atom clusters like elements and dark matter. To return an expanded atom to its original state, it would need to be compressed from all directions. However, even in cases where soft atoms cluster massively in the cores of stars, the pressure from dark matter coming from all directions—due to the dome effect caused by dark matter arrays pressing on each other simultaneously—makes it impossible to compress a photon. Moreover, excessive compression of dark matter can disrupt its own symmetry, leading to a supernova. Similarly, in regions where dark matter is extremely compressed, such as black holes and dark energy zones, it is impossible for a photon to enter. Just as a soft atom cannot revert to its raw, small, solid state, entropy, once initiated, is irreversible.

Conclusion: The fundamental source of entropy is the expanded atom, that is, the soft atom.

61. DARK MATTER AERODYNAMICS IN HELICOPTERS

Aerodynamics is a scientific discipline that studies the interactions of gases—particularly air—with the surfaces of solid objects. It investigates how objects move through the air, how air resistance is generated, and how aerodynamic principles are applied in the design of aircraft, vehicles, or structures. Known as a subfield of fluid dynamics, aerodynamics plays a crucial role in engineering and design processes to enhance the performance of air vehicles (such as airplanes, helicopters, and rockets), vehicles (such as cars and trains), and structures (such as bridges and buildings).

In mainstream science, aerodynamics has been well-explained without considering the dark matter factor. However, according to the Theory of Everything, I will explain aerodynamics by incorporating the dark matter factor. I briefly touched on this topic earlier under the heading "Aerodynamics in Space."

It is commonly thought that a helicopter lifts off by pushing the air above it downward, but the elements in the air are so sparse and small that it seems illogical for them to lift vehicles weighing tons into the air. The primary force behind anything that flies is the force exerted by dark matter. When a helicopter's blades begin to rotate, they push dark matter downward in a fluid motion, much like the propeller of a vehicle moving through water pushes the water behind it. The rapid rotation of the helicopter blades also generates an external magnetic field, increasing the strength of this field. In the dark matter arrays above the newly formed magnetic field, a negative force emerges, while a positive force forms in the arrays below. This creates an invisible reverse vortex (see the Vortex Hypothesis). The altered environment facilitates the helicopter's upward lift. Relative to its own axis, a helicopter can move upward more comfortably and quickly than an airplane, but it cannot move forward as fast as an airplane. This is because the magnetic field created by the helicopter blades resembles a vertical, thick, invisible tube.

You might ask, "Why can't we fly a helicopter in space?" The answer is this: In space, dark matter is so dense relative to elements that the momentum generated by the rotating blades is insufficient to create negative pressure in the dark matter arrays above and positive pressure in the arrays below. As a result, the helicopter cannot move in any direction and simply spins in place.

Additionally, keep this in mind: The atoms on the surface of dark matter on Earth are significantly more deformed compared to those in space.

62. DARK MATTER AERODYNAMICS IN AIRPLANES

According to the Theory of Everything, the angle of a helicopter's blades differs from that of an airplane's propeller. This angle difference is one of the factors that allows an airplane to travel faster. Due to this angle difference, the magnetic field vortex created by the airplane propeller is both sharper and stronger, thanks to the propeller's rotation speed. When an airplane accelerates, it takes off for reasons well-known in mainstream science. Once airborne, it moves quickly due to the pressure generated between dark matter particles. Airplanes glide through dark matter much like a boat skims across the surface of water. However, in space, an airplane cannot move for the same reasons I described in the previous hypothesis (Dark Matter Aerodynamics in Helicopters).

63. DARK MATTER AERODYNAMICS IN ROCKETS

According to the Theory of Everything, the dark matter aerodynamics of rockets and similar vehicles differ from those of airplanes and helicopters. Unlike airplanes and helicopters, as described in their respective hypotheses, rockets do not simultaneously create a vortex between dark matter particles or generate a negative pressure zone in front and a positive pressure zone behind.

Rockets create only a positive pressure zone behind them. During launch, the expelled rocket fuel rapidly ejects elements and photons at the quantum level. These elements immediately break apart, scattering countless electrons into the surroundings. The scattered elements and electrons infiltrate the spaces between dark matter particles, compressing them. In this process, the magnetic field behind the rocket expands, and this expanding magnetic field propels the rocket forward. However, during launch, while the expanding magnetic field is close to the ground, it hinders the dark matter particles in front of the rocket from moving to the rear. As the launch progresses, the dark matter particles in front of the rocket's magnetic field slowly begin to flow toward the rear area. Once the rocket is airborne, momentum mechanics come into play proportionally.

The reason large rockets are slow during takeoff and gradually increase their speed lies in the positive force generated as the dark matter particles in front of the rocket flow to the rear and return to their original positions after the rocket passes. We can also express it this way: For an object to move forward, the dark matter in front must change its position to clear the path, apply pressure from the sides as the object passes, and return to its original position afterward. The intense emission of elements and photons from the rocket pushes the dark matter with a positive force. Consequently, the distance the dark matter in front must travel to clear the path increases. This is because, when large rockets are first ignited, they create a force field on the ground that extends for meters, resembling an inverted mushroom or umbrella. This is why large rockets gain forward momentum slowly. Once the rocket lifts off the ground, the magnetic field shaped like an inverted umbrella closes, and the rocket's speed begins to increase as it ascends.

Rockets can travel in space because the elements emitted from the rocket occupy space between dark matter arrays, creating a positive force among the dark matter particles.

64. TURBULENCE

Aircraft sometimes encounter turbulence; you may have experienced or heard of this phenomenon. According to the Theory of Everything, the cause of turbulence is the sudden change in dark matter pressure when you are in a particular environment or transition to a different one. The effect of turbulence varies depending on whether you move from high pressure to low pressure or from low pressure to high pressure.

For example:

- The impact of falling into the sea from a height of 10 meters is different from the effect of rising to the surface from a depth of 10 meters at a similar speed.
- When driving on certain roads with inclines, you transition from a high-pressure area to a low-pressure area, feeling as if you're flying and experiencing turbulence due to the sudden pressure change.
- When you quickly drive up a mountain from a flat area and then descend rapidly, your eardrums may feel discomfort.

The turbulence experienced by aircraft, which travel at high speeds, is similar to these simple events.

On Earth, there is high pressure in the sea. On land, particularly underground, the pressure is even higher. In the gas-covered terrestrial environment where we live, the pressure is lighter. This lighter pressure results from the sparse presence of elements between dark matter particles and the deformation of dark matter surfaces due to intense photon transitions, as well as the friction of electrons and elements. When you leave Earth, you encounter the highest level of intense pressure. Contrary to common belief, the pressure in space is not millions of times lower but millions of times higher.

This is why radio signals bounce back. Astronauts traveling to space experience turbulence for this reason. Ask astronauts: "What does the turbulence you experience while going to and entering space feel like?"

Note: In the Theory of Everything, there are two types of pressure factors: the positive pressure that dark matter exerts on elements and photons from the outside in, and the positive pressure that elements and photons exert from the inside out. This hypothesis refers to the pressure exerted by dark matter from the outside in.

65. HYPOTHESES ON SOME MEASURING INSTRUMENTS

According to the Theory of Everything, measuring instruments not only quantify known physical phenomena but also enable us to understand the interactions of dark matter, elements, and photons. Below, I will explain the functionality of various measuring devices and how they can be reinterpreted or improved within the framework of this theory.

Barometer

A barometer is a device used to measure atmospheric pressure, often employed in weather forecasting and altitude determination. It comes in two main types: Mercury barometers measure pressure through changes in the level of mercury in a glass tube, while aneroid barometers detect pressure changes using a flexible metal capsule and display them via a pointer. Barometers are widely used in fields such as meteorology, aviation, and maritime activities.

In the Theory of Everything, dark matter consists of clusters of solid atoms, with elements and photons residing between dark matter arrays. On Earth, the surfaces of dark matter are deformed due to friction from the intense movement of elements and photons—while atoms themselves do not wear down, the solid atoms on the surface of dark matter have softened at their tips. Barometers measure the pressure changes in both the atmosphere and dark matter. As you ascend from the ground to the sky, the mercury level or the flexible metal capsule in the barometer reflects these changes. A new device, designed with setups aligned with the Theory of Everything, could be invented to measure and classify dark matter pressure. Such a device could prove highly beneficial in various technological advancements.

Spectrophotometer

A spectrophotometer is a device that measures how materials absorb or reflect light, widely used in chemistry, biology, and physics. By measuring a sample's light absorption at different wavelengths, it helps determine the sample's concentration and composition. It typically operates in the ultraviolet (UV), visible, and infrared (IR) regions. For instance, the concentration of a solution can be analyzed using the Lambert-Beer Law. Spectrophotometry has a broad range of applications, from medical diagnostics to environmental analysis.

In the Theory of Everything, when a photon travels between dark matter particles and encounters a single element, it loses some of its components depending on the element's size and continues its journey by passing around the element. By developing a larger and more sensitive spectrophotometer, elements can be reclassified. Since the atom model in this theory differs from conventional ones, this approach would allow for a more detailed classification of elements.

Thermometer

A thermometer is a device used to measure temperature. While there are different types, they all rely on a material sensitive to temperature changes. In mercury thermometers, as the temperature rises, mercury expands, raising its level in a glass tube. Digital thermometers measure changes in electrical resistance using a thermistor or thermocouple, converting them into temperature readings. Infrared thermometers determine temperature by detecting the infrared radiation emitted by objects. Thermometers are vital in fields like healthcare, meteorology, industry, and scientific research.

According to the Theory of Everything, negative pressure in dark matter arrays creates a hot environment, while positive pressure creates a cold one; the formation of one necessitates the other. Measurements conducted with hundreds of thermometers in an isolated environment can help us analyze the behavior of dark matter arrays.

Hygrometer

A hygrometer is a device used to measure humidity in the air, which refers to the amount of water vapor present. It comes in two main types: Mechanical hygrometers operate based on the expansion and contraction of organic materials (such as human hair) with humidity, while electronic hygrometers measure humidity by detecting changes in capacitance or resistance. Hygrometers are commonly used in weather forecasting, monitoring greenhouses and storage facilities, museums, and other areas where humidity control is critical.

In the Theory of Everything, if hundreds of hygrometers are arranged at symmetrical angles in an isolated environment, mathematical models about the fluid dynamics of gases can be developed.

Anemometer

An anemometer is a device used to measure wind speed and often its direction, widely utilized in weather stations, forecasting, and research.

According to the Theory of Everything, dark matter is not typically fluid, but extreme pressure changes can cause dark matter to flow with gases to a limited extent. To determine the pressure under which this flow occurs, experiments can be conducted with hundreds of anemometers in an isolated environment. This could enable the prediction of powerful natural events like hurricanes.

Pycnometer

A pycnometer is a laboratory device used to precisely measure the density of liquids. Typically made of glass, it consists of a closed container with a specific volume and a lid. The pycnometer calculates density by dividing the mass of a liquid by its known volume.

In the Theory of Everything, mathematicians have a role to play: "Can a pycnometer help calculate how many dark matter particles, detached from their arrays, are present in a given type of liquid?"

This question could be a significant step in understanding the dark matter content of liquids.

Manometer

A manometer is a device used to measure the pressure of gases or liquids. Although there are different types, they share the same basic principle: They measure pressure by observing changes in the height of a liquid column or the deformation of an elastic element.

In the Theory of Everything, the working principle of a mercury manometer can be used to analyze the contribution of dark matter particles—those that detach from their arrays to enable the liquid state of matter—to the positive force exerted from the inside out after transitioning from gas to liquid.

Voltmeter

A voltmeter is a device used to measure the potential difference, or voltage, between two points in an electrical circuit, commonly used in electrical engineering, electronics, and physics laboratories.

According to the Theory of Everything, a voltmeter actually measures the single fundamental force in nature—in other words, the strength of the positive force experienced by electrons during their flow. By passing a laser beam or electricity through tunnels or cables designed like a maze with adjustable walls, voltmeter measurements can be analyzed. This setup would allow for the study of photon behavior in response to magnetic fields within the tunnels, as well as other analyses from various perspectives.

Ammeter

An ammeter is a device used to measure the electrical current in a circuit. Current refers to the flow of electrical charges from one point to another, and ammeters typically measure this in amperes (A).

In the Theory of Everything, the working principle of an ammeter can be adapted to a setup where water is sent through pipes. In this arrangement, the water pipes represent conductive cables, and the interior of the pipes should be designed to mimic elements—meaning the pipes' interiors should not be smooth. By sending water through these pipes, the molecular arrangements of conductive and insulating materials can be simulated. The measurements obtained with water can then be correlated with existing ammeter measurements.

66. IS DARK MATTER THE SAME AS ETHER?

The term "ether/aether theory" refers to a philosophical and scientific concept, often rooted in aristocratic and materialist philosophies. According to this theory, there is a relationship between matter and energy (ether). Ether is considered an invisible, universal substance that serves as the medium enabling the existence of matter. Since ancient times, the concept of ether has appeared in philosophical and scientific discussions, though its meaning has varied over time.

However, in the Theory of Everything, dark matter is neither ether nor energy. First and foremost, matter is the cause, while energy is the result. To put it metaphorically, dark matter is not ether but a cage.

The Theory of Everything progresses through inductive reasoning, starting from infinite singularity. Through this process, a space enveloped by dark matter has formed, and this structure has continued to evolve, giving rise to everything. Before developing this theory, my knowledge of physics was as limited as that of the average person. I learned everything I know about mainstream physics later on. Using this acquired knowledge, I associated the components of the Theory of Everything—such as dark matter, elements, photons, and electrons—with concepts in mainstream physics.

Everything I have defined has countless physical proofs already in existence. The concept known as ether can be correlated in the Theory of Everything with the energy that emerges from the pressure exerted by symmetrical dark matter and its arrays on the asymmetrical elements and photons between them, as well as the counterforce these elements and photons exert in response.

67. THE MICHELSON-MORLEY EXPERIMENT

The Michelson-Morley Experiment, conducted in 1887 by Albert A. Michelson and Edward W. Morley, is a significant milestone in scientific history. The experiment aimed to test the existence of a hypothetical medium called the "ether," which was thought to carry light.

At the time, the existence of the ether was widely accepted, and it was believed that light propagated through this medium. If the ether truly existed, the Earth's motion through it should have caused the speed of light to increase or decrease in certain directions. The experiment sought to measure whether the Earth's movement affected the speed of light within the ether. However, Michelson and Morley did not obtain the expected results. Their experiment demonstrated that the speed of light remained constant across different seasons and directions of the Earth's motion. This finding seriously challenged the ether theory and paved the way for Albert Einstein's development of the theory of special relativity. Special relativity proposed that the speed of light is constant in any reference frame, eliminating the need for an ether medium. The Michelson-Morley Experiment is regarded as a crucial turning point in the development of modern physics and the abandonment of the ether concept in scientific thought.

According to the Theory of Everything, all living beings and objects exist between dark matter arrays. When any observed object or living entity forms within these dark matter arrays, it first creates a magnetic field and then a cage wall. This magnetic field and cage wall are strong relative to the positive force exerted from the inside out by the dark matter, elements, and/or photons trapped within. Therefore, the Michelson-Morley Experiment was conducted within a magnetic field. Inside the cage wall of a magnetic field, dark matter arrays do not flow. In reality, it is the objects and living beings that move within dark matter arrays. However, dark matter within the vortices created by planets and stars does flow. For example, within the Sun's vortex, each planet has its own distinct vortex. Consequently, to obtain reliable results from the Michelson-Morley Experiment, it should be conducted at the boundary between Earth and space or at the edges of planetary vortices. Additionally, dark matter entering black holes transforms into dark energy and exits from the other side; this can be considered a relative flow.

68. LIGO (Laser Interferometer Gravitational-Wave Observatory)

LIGO (Laser Interferometer Gravitational-Wave Observatory) is a large-scale scientific facility designed to directly observe gravitational waves.

According to the Theory of Everything, instead of building massive facilities like LIGO, smaller, 100-square-meter miniature LIGO facilities could be established worldwide on solid surfaces—such as coastal areas at sea level and mountain peaks. By analyzing data from these facilities, the dark matter pressure force resulting from the Sun's solar flares and the geometric alignments of the Moon and planets could be measured. Furthermore, scientific studies spanning a wide range of phenomena—from atmospheric changes to major earthquakes—could be conducted. Establishing numerous smaller LIGO-like facilities across diverse geographical locations could be highly beneficial for understanding the influence of dark matter on the fluid dynamics of gases.

69. CERN

CERN (European Organization for Nuclear Research) is one of the world's largest particle physics laboratories. Its primary goal is to study and understand fundamental particles and their interactions by conducting high-energy particle collisions.

According to the Theory of Everything, all structural properties of an atom can be explained solely through its solid and soft states. In this context, what else could CERN be used for? For instance, it could continue colliding elements to determine how many atoms each classified element consists of.

70. DESERTS

Deserts are among the driest and most challenging environments on Earth. These regions, characterized by vast expanses of sand or rock, receive extremely low annual rainfall. Temperature differences can be drastic within a single day, with scorching heat during the day and freezing cold at night. Due to these conditions, plant cover and animal diversity are highly limited.

According to the Theory of Everything, the scarcity or complete absence of plant cover in deserts is due to dark matter exerting negative pressure parallel to the Earth's crust. In areas with negative pressure, the surfaces of dark matter and elements have a high concentration of electrons. This causes desert soil to have a loose structure parallel to the Earth's surface. In such a loose environment, there is no supportive force to help trees grow upward or grasses grow downward. Without this force, condensation in the air is limited, leading to very little rainfall and preventing the formation of plant cover.

71. SNOWFLAKES

Snowflakes and snow crystals are among the most enchanting natural phenomena of winter. Snowflakes form when water vapor in the atmosphere freezes into tiny ice crystals. Each snowflake develops unique and intricate hexagonal patterns, which vary depending on atmospheric conditions such as temperature and humidity. While snow crystals are typically too small to be seen with the naked eye, their detailed and symmetrical structures become visible under a microscope. Snowflakes can take various forms, such as stars, columns, or needles. The uniqueness of each flake reflects the diversity and complexity of nature. Beyond their aesthetic appeal, snow plays a significant role in the climate and water cycle. Snow cover reflects sunlight, influencing the Earth's temperature balance, and melts to replenish water sources. For these reasons, snowflakes and snow crystals captivate both scientific and artistic interest.

According to the Theory of Everything, the entire universe is enveloped by dark matter, meaning that dark matter exists both inside and outside every living being and object—except for nuclear bombs and massive photons, which contain no dark matter. Snow crystals may serve as one of the objects that demonstrate the presence of dark matter through their shapes.

When elements are sparsely distributed among dark matter arrays, matter exists in a gaseous state; the air we breathe is an example of this. The sparsely distributed elements in the air provide a clear sky. However, when external elements are introduced to these sparse elements, clouds form in the clear sky, and the gas transitions into a visible gaseous form. On a cloudy day, if dark matter is compressed by factors located kilometers away, the elements between the dark matter arrays are naturally compressed as well. These compressed elements envelop the dark matter arrays, detaching them from their arrays, and form their own magnetic fields, falling as rain in the form of water droplets.

When dark matter undergoes even greater compression, a different phenomenon occurs. The compressed dark matter further compresses the elements between them. These elements begin to cluster around a larger central element, adhering to it. Starting from the center, the elements envelop the dark matter, forming crystals. These crystals instantly increase their magnetic field and weight, causing them to fall to the ground. Since dark matter often has hexagonal surfaces, snowflakes, which form rapidly between dark matter arrays, typically take on hexagonal shapes as well. The shapes of dark matter play a guiding role in the formation of many structures in nature.

72. TIME OR CHANGE IN SPACE?

The concept of time functioning differently in space is explained by Albert Einstein's theories of general and special relativity. According to general relativity, time flows more slowly near massive objects. As a result, clocks near the Earth's surface run more slowly compared to those farther away in space; this phenomenon is known as "time dilation." Special relativity, on the other hand, states that the time experienced by moving objects flows more slowly relative to stationary ones, with this effect becoming more pronounced at high speeds, particularly near the speed of light. In space travel and for satellites in orbit, these effects are accounted for to adjust time measurements. For instance, the clocks on GPS satellites are calibrated to account for these relativistic effects to remain synchronized with clocks on Earth, ensuring accurate and precise location data. Relativity theories enable us to understand the concept of time in space and optimize technological systems.

However, in a scientific framework based on multiple theories, it's not surprising to find significant misinformation and fallacies. According to the Theory of Everything, whether a person is in space, underwater, or at the event horizon of a black hole, as long as their consciousness is active, time is determined by the brain's algorithmic processing speed and capacity. Since time is formed in the brain, it is metaphysical, whereas changes perceived through the senses are physical. Even if a person's five sensory organs were impaired or nonfunctional, they would still be aware of time as long as they remained conscious.

As for the claim that "time flows differently in space for clocks and devices," this is a fallacy. In objects without a brain, there is no such thing as time; there are only reactive changes in response to stimuli. As long as the stimulus remains unchanged, the reaction does not change either. A mechanical clock on the Earth's surface and one in space will, of course, operate differently because the levels of external forces in these environments differ.

Except for nuclear bombs and massive photons, all entities—whether naturally formed living beings or objects, or human-made tools and devices—contain clustered dark matter. This dark matter becomes detached from its arrays during the formation or manufacturing process, and its surfaces become deformed—that is, expanded, swollen, or softened. Such tools and devices do not function or undergo the same changes in the same way on Earth and in space. The reason for this is that the external-to-internal dark matter pressure is greater in space. When the dark matter pressure force increases, the soft atoms and elements on the surface of the dark matter within the object become compressed. This compression restricts their vibrations, similar to the hand-and-fly analogy. A weaker version of this phenomenon occurs inside rapidly moving vehicles. As you ascend through Earth's atmosphere, dark matter pressure increases. In a fast-moving airplane, the dark matter pressure from the outside increases due to acceleration, and this heightened pressure restricts the vibrations of the elements inside the plane.

73. SEASONAL FLU

Seasonal flu is a viral infection that affects the respiratory system, with its history dating back to the earliest days of humanity. In this hypothesis, I will explain the external factor mechanics behind the flu and other respiratory diseases.

According to the Theory of Everything, every human—whether in the womb or in nature—reaches adulthood through a single cell absorbing external elements and photons. Whether in the womb or in nature, the natural selection and gender of this single cell are determined by the structure and arrangement of elements in solid-state viruses. Moreover, the elements within a single cell are proportionally balanced in terms of percentages. When this balance is disrupted, depending on the severity, outcomes such as birth defects, illness, or death may occur.

To make this clearer, let me provide an example using millimeters: A person's genetic makeup consists of elements ranging from 0 to 10 millimeters in diameter, and these elements are proportionally balanced in number. For instance, if the proportion of 2-millimeter elements increases in the blood, adrenaline levels rise. These 2-millimeter elements can come from external sources or be secreted by the brain. Let's assume that a high concentration of 9- to 10-millimeter elements leads to flu symptoms.

When winter arrives, the dark matter pressure in the air increases, and the photon bombardment from the Sun decreases compared to summer. As dark matter pressure rises, the vibrations of elements in the air diminish. Elements with reduced vibration levels combine with other elements, transforming into viruses that remain suspended in the air for longer periods. When people breathe this air, they may also inhale these viruses. If the inhaled viruses disrupt the elemental balance in your genetic makeup, you become ill. This phenomenon is less common in summer compared to winter, which is why fewer people get sick during the warmer months.

74. VACCINES

According to the Theory of Everything, if living beings, after birth, continue to be nourished proportionally with elements and photons in accordance with the genetic structure they possess as a single cell, they will reach adulthood in a healthy manner. Vaccines are element supplements administered to healthy individuals. The content of vaccines, especially when given at a young age, can alter the genetic structure of all living beings.

In our theory, a vaccine is a genetic intervention performed on a healthy organism. In humans, everyone's constitution varies slightly. Applying mass-produced, one-size-fits-all vaccines to people with these varying constitutions may be ineffective for most, while being either harmful or beneficial to a small number of individuals.

The best approach is to obtain elements through natural foods during regular nourishment rather than through mass-produced artificial vaccines. This is because foods compatible with an individual's genetic structure can perform functions similar to those of artificial vaccines.

75. MEDICINES

According to the Theory of Everything, medicines are tools used to treat illness by either supplementing or removing an element that disrupts the percentage balance of elements in the genetic structure. Alternatively, instead of expelling an element whose proportion has increased, balancing the ratio by increasing the number of other elements in the body is also considered a form of medicine. When this balance is achieved, the immune system becomes stronger against the specific disease.

The key to staying healthy for all living beings is singular: maintaining the percentage balance of the elements that form your genetic structure when you are a single cell. The foods we consume can alter our genetic structure on a micro level. As long as this change is not excessive, it can be beneficial rather than harmful. Whether it's cancer or a cold, the principle remains the same—you must restore the percentage balance. If you have taken in an element larger than those constituting your genetic structure and this has led to cancer, you must either remove that larger element through surgical intervention or reduce its size by increasing its electron emission. In short, the percentage balance and volumetric size of elements must be maintained in equilibrium.

76. IMMUNE SYSTEM

According to the Theory of Everything, since everything originates from three main classes, all systems are inherently simple. The natural functioning of what is defined as the immune system is also remarkably straightforward.

The foundation of the immune system, from galaxies to bacteria, lies in the strength of the magnetic field and its cage wall. It is primarily this magnetic field cage wall that protects us from harmful elements and photons. Within our body, each organ also has its own unique magnetic field cage wall. This structure is the first element that prevents our body from disintegrating and dissolving like water.

The second crucial component of the immune system is the digestive system—the process through which everything, including the air we breathe through our mouth and nose, passes. The digestive system acts like a filter. The things we consume attempt to pass through these filters. If harmful, larger elements pass through the filter's pores, we may become ill or even develop cancer. Smaller elements, however, pass through the filter and turn into blood that nourishes our organs. For example, the organ where air is "digested" is the lungs. The air we breathe is filtered in the lungs' sieve; oxygen passes through the tiny pores and enters the bloodstream, while elements that get trapped and cannot pass are expelled as carbon dioxide.

The third key component of the immune system is the pool of elements that form our genetic structure. The wider a person's asymmetrical element pool, the stronger their immunity. This element pool expands when we fully recover from diseases we've experienced. This is the primary purpose of vaccines administered in infancy—they aim to expand the baby's genetic pool. However, since vaccines are mass-produced, they may not be compatible with every baby's genetic pool, potentially causing side effects or more serious harm. I hope that in the future, vaccination will be tailored to individuals' genetic pools, similar to how blood types are classified. Nevertheless, I must emphasize that the healthiest form of "vaccination" comes from food. Elements obtained from food are filtered by the body before being processed. Medical vaccines, on the other hand, are directly injected, bypassing the digestive system, which makes them quite risky and, in some cases, can endanger a person's life.

77. GRAVITATIONAL FORCE

According to the Theory of Everything, there is no such thing as a gravitational force. The phenomenon observed between two masses arises from the vector force directions of dark matter arrays.

The orbiting of planets is determined by the vortex created by a star. The paths or orbits of satellites and smaller objects are shaped by the unique vortices each planet generates. Satellites with a photon core that have not yet faded may also have their own vortices. In a star system formed by vortices, the three-body problem does not occur.

The falling of objects to the ground is as I explained in the Weight Hypothesis.

78. ELECTROMAGNETIC FORCE

According to the Theory of Everything, the electromagnetic force exists, but it arises from the reactive vectors of symmetrical dark matter, which are automatically generated by the influence of the fundamental inside-out positive force.

The carrier particle of this force is the photon, which possesses a weak inside-out positive force. Electricity and magnetism affect chemical bonds as defined in mainstream science.

79. STRONG NUCLEAR FORCE

According to the Theory of Everything, since the atom model is different, there is no such concept as the strong nuclear force.

In our theory, an atom is an indivisible particle. What keeps a solid atom stable is the stability of the other atom it is in contact with. Events defined as nuclear fusion and fission occur within elements. An atom itself cannot be split, but elements, which are clusters of solid atoms, can be split.

80. WEAK NUCLEAR FORCE

According to the Theory of Everything, since the atom model is different, there is no such concept as the "weak nuclear force."

The phenomenon known as radioactive decay is similar to the hand-and-fly analogy: When an element is subjected to negative pressure from surrounding dark matter, the solid atoms on its surface expand, turning into electrons, and move away from the source element. During this process, the element loses mass.

The Sun's energy production occurs through the emission of electrons from elements. During explosions on the Sun, some dark matter detaches from its arrays and breaks apart, resulting in the formation of new elements from external sources.

81. PETROLEUM AND MICRO GRAVITY

According to the Theory of Everything, elements are asymmetrical. This asymmetry leads to a wide variety of shapes; elements can take countless forms such as flat, thin and long, triangular prisms, or cubic prisms. The asymmetrical shape diversity of elements sometimes plays a significant role in clustering. In mainstream science, the stability of atoms or what is referred to as "micro gravity" may be a result of this phenomenon.

The clustering of elements by class is quite simple. For example, if one triangular prism is placed among 10 cube-shaped materials, vibrations and external pressure will cause the cube-shaped materials to unite and form a new cluster, excluding the triangular prism. Symmetrical shapes tend to align side by side or nest within each other. If symmetrical shapes are in the majority, they exclude asymmetrical shapes, or vice versa.

The elemental structures of materials like petroleum, gold, and silver enable them to form new clusters. For instance, petroleum, due to its shape and mass, seeps through other asymmetrical elements and clusters underground. This is the reason behind the formation of natural reserves, deposits, and veins. A similar clustering mechanism is at play in the formation of human organs.

Conclusion: There is no such thing as micro gravity or gravitational force. These phenomena occur due to materials excluding each other based on their shapes and/or volumetric sizes. Symmetrical entities push out asymmetrical ones, and asymmetrical entities push out symmetrical ones from the environment.

82. POPCORN

According to the Theory of Everything, at the micro level, there are no concepts of hot or cold; there are only the solid (compressed) state and the soft (loose) state of an atom. At the macro level, heat forms in an environment with negative pressure in dark matter arrays, while cold forms in an environment with positive pressure. It would be helpful to keep the hand-and-fly analogy in mind.

When a photon bombardment is applied to an element or a cluster of elements, the cluster expands from the inside out due to the external photons. This expansion creates a negative pressure zone in the dark matter arrays outside the cluster's surface. This phenomenon is similar to the loosening of your fist in the hand-and-fly analogy, but in this case, the fly pushes the hand due to the photon bombardment. In other words, photons momentarily overpower the inside-out positive force. When a positive pressure zone forms in one area, a negative pressure zone forms simultaneously in another area, or vice versa.

Have you ever eaten popcorn or popped corn yourself? A small corn kernel, in terms of volume, becomes significantly larger when it pops. How does this phenomenon occur?

The outer shell of a corn kernel is hard, while its interior is soft. When you apply a photon/electron bombardment to pop the corn, countless solid atoms on the surface of the elements forming the corn kernel turn into soft atoms, resembling tiny bubbles. As a result of this transformation, the interior of the corn kernel breaks through its outer shell. From the broken shell, numerous lightweight elements rapidly escape, slipping through the surrounding dark matter arrays and incorporating many dark matter particles into the cluster. Consequently, the corn kernel expands and becomes lighter relative to its increased volume. As I explained in the Weight Hypothesis, the weight of an object is directly related to the area it occupies and the proportional space taken up by elements and dark matter within that area.

83. EGG

How can the inside of an egg, which is a fluid liquid, transition to a solid state when heated, while most liquids, including water, only become solid when frozen?

According to the Theory of Everything, the inside of an egg is composed of small and lightweight elements that have been synthesized and bonded together, while its shell is made of slightly larger elements. When the egg is boiled in water, its elements continuously vibrate. These vibrating elements stretch the solid atoms on each other's surfaces like elastic, causing them to expand. As countless elements swell by being coated with electrons, they more tightly envelop the dark matter between them, leading the egg to transition from a liquid form to a semi-solid form as it heats.

84. BOILING AND BURNING

According to the Theory of Everything, boiling and burning cause different reactions in materials. When a material is boiled, in addition to its existing magnetic field cage, it is subjected to the positive pressure of the surrounding liquid, which reduces element loss. This reduction in element loss ensures that the nutritional value of boiled food is higher than that of food cooked by burning.

In contrast, materials that are burned experience a photon bombardment from the fire, which causes a greater loss of elements compared to boiling.

85. CABLES, CRYSTALS, AND GLASS

According to the Theory of Everything, nothing can pass directly through another thing. In a cable with an electric current, the current flows as electrons move between elements in a manner similar to the flow or slip of water. When electrons exit the magnetic field cage wall, they separate from the cluster and transform into photons. However, electrons cannot travel parallel to the cable along the surface of the cable cluster.

The situation is similar in crystals and glass. Photons cannot pass directly through a crystal or glass, nor can they detour around its magnetic field to continue on the other side. When a photon strikes a crystal or glass, it transfers its kinetic energy to the electron it hits. At the moment of impact, the crystal's magnetic field cage wall opens and closes while applying a positive force. Due to the combined force from dark matter and the photon's speed, for every photon element that transforms into an electron, an equal number of electrons are pushed out from the other side, transforming back into photons.

86. SOAP BUBBLES

Creating bubbles from soap is a fun and simple activity. A soap solution is prepared, and bubbles of various sizes are formed by blowing through a bubble wand. Have you ever thought about how this works in terms of quantum mechanics?

According to the Theory of Everything, when you dip a bubble wand into the solution and pull it out, the resulting film is so thin that its magnetic field cage wall is almost nonexistent. If you blow quickly through the wand, the film breaks immediately. However, when you blow slowly, the elements forming the film begin to slide proportionally between dark matter arrays, while the dark matter incorporated into the film's cluster does not move. The elements release the dark matter they had trapped, allowing the bubble to expand as it inflates. Eventually, the expanding film detaches from the bubble wand. At the moment of detachment, the thin film separates the dark matter trapped between the elements from their arrays, enabling the bubble to float away.

On the surface of the bubble's film, a vibrant array of dancing colors appears. At the same time, however, the momentum of the photons in the photon bombardment creating these colors reduces the mass of the elements forming the film. When the bonds of these mass-reduced elements weaken sufficiently, the bubble bursts, as the magnetic field cage wall of the film is extremely fragile.

As can be seen, in objects with a weak magnetic field cage wall, elements can move forward by releasing the dark matter they had trapped and capturing other dark matter. In contrast, objects like iron or water droplets, as well as all living beings, move along with the dark matter they trapped at the moment of their formation.

The soap bubble phenomenon demonstrates that elements can progress by moving between dark matter arrays, transitioning from one dark matter particle to another.

87. SHADOW

A shadow is a dark area formed when light from a source is blocked by an object. The size and shape of a shadow vary depending on the position of the light source and the shape of the object. Shadows are an easily understood phenomenon in classical physics fields like optics and geometry. However, from a quantum mechanics perspective, shadows become more complex.

According to the Theory of Everything, at the quantum level, no photon can travel in a straight line at any wavelength due to the shape of dark matter. Even a laser beam, which we assume travels straight on a macro scale, vibrates as it moves at the quantum level. As a photon travels, it constantly loses elements in all directions. It compensates for these lost elements by absorbing a limited number of elements lost by other photons. However, a photon loses the most elements when it collides with an object or a living being.

When a cluster of photons collides with a cluster of elements, it breaks apart, scattering countless elements in all directions. At the moment of impact, depending on the element structure of the cluster and the strength of its magnetic field wall, some photon components transform into electrons, pushing an equal number of electrons out from the other side. These expelled electrons then transform back into photons, but these photons are no longer in the visible light wavelength. In clusters of opaque elements, a shadow forms because the number of photon elements significantly decreases, and their wavelength lengthens.

Photons that do not collide with the object but pass nearby compress the dark matter in the area behind the object where the shadow forms as they move through the dark matter arrays. This compressed dark matter causes the shadowed area to be cooler. This is why shadows are cool and why nights are dark and cold.

88. PIXELS

A pixel is the smallest building block of digital images, derived from the term "Picture Element." Each pixel in an image has a specific color and brightness. By combining pixels, digital photos, videos, and screen displays are created. Screen resolution is determined by the number of pixels; a higher number of pixels results in clearer and more detailed images.

According to the Theory of Everything, each element in a digital screen emits electrons, which arrive via electricity, as photons outward from the screen based on its positional state. This process occurs through the positional changes of elements, triggered by algorithmic operations at the quantum level. You can think of the elements forming pixels as turnstiles at a gate. For example, an element acting as a turnstile might allow 10 electrons to pass when positioned horizontally, producing a yellow color, while allowing 20 electrons to pass when positioned diagonally, resulting in a different color.

89. TOUCHSCREEN

Touchscreens, commonly used in mobile phones, typically operate using capacitive technology. Capacitive screens work by detecting the electrical charge of the human body. The screen is coated with a special material designed to sense changes in electrical charge. These changes are detected by sensors beneath the screen, which identify the points of touch.

According to the Theory of Everything, everything can be explained. Recall the turnstile analogy from the previous hypothesis! The images we see on digital screens are the result of algorithmic operations formed by 1s and 0s. These algorithmic operations occur through finite physical interactions and reactions at the quantum level, happening in a chain-like sequence. The 1s and 0s represent the moments when electron flow is either enabled or interrupted.

When you touch a touchscreen, the electrons in your finger push against the electrons on the screen, initiating an algorithmic operation. In other words, your finger performs the function of a hardware component inside the computer. For example, when you press and hold your finger on the phone screen, it sends a code sequence like 1, 1, 1, 1, 1...; when you tap and release, it sends a 1, 0 code. The tasks performed by a mouse cursor on a computer are carried out by your finger on a touchscreen.

90. OUR EYES AND NATURE'S PIXELS

According to the Theory of Everything, the visual processing mechanism of our brain differs from the previously discussed topics of pixels and touchscreens only in its natural nature. Every pixel we see in nature through our eyes is composed of photons originating or reflecting from elements. Photons always move as particles, but the rapid and numerous movements of photons create a wave function.

Each photon passing through our eye corresponds to a 1 in computer coding, while the interval between two photons represents a 0. The magnetic field cage wall, made of dark matter and protecting our eye, can be likened to the semiconductors found in computers. The magnetic field and cage wall in our eyes protect us from harmful, large photon clusters while also preventing very small photons from passing through. This is why the concepts of visible light wavelengths and invisible light wavelengths have been defined.

The magnetic field of the eyes also enables the perception of colors. This is why animals and humans perceive colors differently. Each photon from a pixel passing through our eye triggers the rod cells, which act as detectors, initiating algorithmic operations in our brain.

91. CLOUDS AND WOOD FIRE

Imagine you've observed a piece of wood burning—whether in a fireplace, a campfire, or while cooking. The flames from the wood fire rise, disappear, and rise again. As you watch the fire for a prolonged period, you might notice a rhythm in these rises. You may even observe that the flames don't rise uniformly everywhere. Similarly, you've likely observed clouds in the sky, particularly those that look like fluffy, scattered cotton.

According to the Theory of Everything, both clouds and wood fire occupy space among dark matter, creating an inside-out positive pressure in their respective areas. In other words, the elements and dark matter in the regions where fire and clouds exist expand spatially. This expansion compresses the surrounding dark matter. As previously mentioned, if a positive pressure zone forms in one area of dark matter, a negative pressure zone forms nearby.

This is why, where there is heat in one place, there is a cooler area nearby. If photon movement is intense in one region, it becomes sparse in a nearby region. When the flame rises at one point in a wood fire, it diminishes at another point for this reason. Scattered cloud clusters have an irregular, jagged shape due to this phenomenon. The indentations and protrusions of continents fit together for the same reason. Sunspots form for this reason as well. This could even be defined as a law of physics.

92. FLUID DYNAMICS OF GASES

The fluid dynamics of gases is a scientific field that studies the motion and interactions of gases. This discipline is critically important in applications across engineering, meteorology, aviation, and space research. Physical properties such as the density, pressure, and temperature of gases form the foundation of fluid dynamics. Understanding gas flow has wide-ranging applications, from designing jet engines to predicting weather patterns. Phenomena like gas compressibility, turbulence, and shock waves are among the deeply studied topics in this field.

According to the Theory of Everything, what we call "gas" is an area formed by elements sparsely distributed among dark matter. The behavioral mechanics of gases depend on the interactions between the negative and positive pressure zones of dark matter.

Imagine you've blown tobacco smoke in a closed room or seen someone do so. The moment the smoke is blown, the elements in the smoke exert an inside-out positive force among the dark matter. This positive force simultaneously creates negative pressure zones, and this event continues as a short-lived chain reaction. You can observe this dynamic when watching the movement of visible gases. The dynamics of gases are determined by the dance of forces within dark matter arrays.

In large gas masses like Earth's atmosphere, the effects of dark matter are observed over broader areas. For example, influenced by the Moon, during natural events such as volcanic eruptions, large fires, or hurricanes, dark matter can trap elements in the area, forming clusters. In such cases, the elements and dark matter within the gas cluster move together. In other words, a flow occurs among dark matter with dark matter. These types of flows can lift trucks into the air or tear off and blow away the roofs of houses.

Conclusion: In light breezes or windless conditions, only the elements move within dark matter arrays. However, in situations with strong winds, elements flow together with dark matter, moving collectively among dark matter arrays.

93. FLUID DYNAMICS OF LIQUIDS

The fluid dynamics of liquids differ significantly from those of gases. In gases, elements are sparsely distributed among dark matter, whereas in liquids, elements are far more densely packed. In liquids like water, countless elements detach numerous dark matter particles from their arrays, surrounding and trapping them. Recall the bearing analogy mentioned earlier: we likened the balls to elements and the inner ring of the bearing to dark matter. Imagine this in three dimensions, picturing all the water on Earth. Countless dark matter particles are trapped within water. The symmetrical arrays of dark matter exist only outside the water.

According to the Theory of Everything, elements are asymmetrical. For example, freshwater—the water we drink—is composed of lightweight elements. Seawater also contains lightweight elements, but it has a significant amount of heavier elements as well, which is why it is salty. Let's say the water we drink consists of elements with a diameter of 0–2 millimeters, while saltwater consists of elements with a diameter of 0–4 millimeters. This volumetric asymmetry prevents freshwater and saltwater from mixing immediately. The reason is that the magnetic field cage wall of saltwater is stronger than that of freshwater.

The clusters formed by these asymmetrical element arrangements within water are the cause of underwater and surface currents in the ocean. On land, the flow of liquids is entirely related to the positive pressure force exerted by the dark matter arrays outside the water. As mentioned in previous hypotheses, the pressure force of dark matter arrays increases as you move upward from the ground to the sky. Like countless other entities, including photons, water flows from areas where dark matter pressure is stronger to areas where it is weaker. This is why water tends to flow downward from higher to lower elevations. In places like space stations, where the external dark matter pressure is approximately equal, water droplets remain suspended in the air.

94. GULF STREAM CURRENT

The Gulf Stream is a powerful, warm, and swift ocean current in the North Atlantic Ocean. It originates in the Gulf of Mexico, flows along the eastern coast of North America, and continues toward Northern Europe. This current transports a significant amount of heat from equatorial regions to colder northern latitudes, playing a crucial role in regulating climate. The Gulf Stream keeps Western Europe's coasts temperate and acts as a stabilizing factor in the global climate system. Additionally, it serves as an important route for maritime transport and activities.

According to the Theory of Everything, the magnetic field cage walls of warm and cold water have different strengths. The elements in warm water vibrate more than those in cold water. These vibrating elements emit electrons, causing the area occupied by warm water to expand. However, this expanding area is suppressed by the surrounding vast cold water mass, preventing the electrons in warm water from activating those in cold water. When electrons cannot move sideways, they are directed toward the water's surface.

When warm water fails to heat the cold water, an event occurs similar to the hand-and-wet-soap analogy. The positive force exerted by the cold water on the warm water pushes the warm water, resulting in the formation of the Gulf Stream current. The Gulf Stream and similar currents persist as long as Earth's distance from the Sun remains unchanged or the equatorial region is not covered by dust clouds from a volcanic eruption. However, in the case of an extreme natural disaster, their route may shift, or their strength may diminish.

95. ELECTRIC EEL

The electric eel, a fascinating fish species capable of generating electricity, inhabits the freshwater regions of South America. Scientifically named Electrophorus electricus, this fish produces high-voltage electrical signals for hunting, defense, and communication. Its electric organs consist of thousands of specialized electrocyte cells, which together generate an electric potential. The electric eel's ability to produce electricity has inspired research in biophysics and neurophysiology. Furthermore, these fish play a significant role in studying biological energy conversion mechanisms and developing biomimetic technologies.

According to the Theory of Everything, everything is simple. If a living organism can emit electricity outward, it means it can release the electrons it produces and stores. Electrons are a naturally occurring component in all organic life. In our theory, electron production occurs only in areas with negative pressure. If this fish generates electricity, it must be creating negative pressure zones within its body. This process, similar to the hand-and-fly analogy, can be achieved through muscle movements.

To accomplish this, the fish accumulates elements from the food it consumes in a specific area of its body. It then creates negative pressure to produce electrons, which are subsequently expelled outward using a positive pressure force.

96. POLAR LIGHTS

Polar lights, also known as auroras, are mesmerizing light displays visible in Earth's polar regions. This natural phenomenon occurs when charged particles from the Sun interact with Earth's magnetic field. In the Northern Hemisphere, they are called Aurora Borealis (Northern Lights), and in the Southern Hemisphere, they are known as Aurora Australis (Southern Lights). The charged particles collide with oxygen and nitrogen molecules in the atmosphere, emitting light in various colors, typically green, red, purple, and blue. Polar lights are a significant subject of scientific research and a major attraction for tourists.

According to the Theory of Everything, wherever there is a light or heat source in space, a negative pressure zone has formed. First, the only entities that can reach Earth from the Sun are photons. Photons do not have negative or positive charges; they are clusters of soft atoms. The concept of "+" and "–" charges in photons or electrons may arise from the expansion caused by the inside-out positive force they generate while moving through magnetic fields formed by dark matter.

In our theory, photons from the Sun create a bulge, or an inside-out positive force, in the dark matter arrays on Earth's daytime side. This positive force is so strong that it expands all dark matter arrays in the east, west, north, and south directions. The expanded area exerts positive pressure on the surrounding dark matter around Earth, causing those areas to become cooler. The poles, however, are the regions subjected to the greatest pressure due to Earth's rotation on its axis. This is the primary reason why Earth is flattened at the poles and why the poles are cold.

Due to the pressure at the poles, dark matter becomes more compressed. As dark matter density increases toward the poles, small negative pressure zones form in the transition regions. In these negative pressure zones, elements already present in the air emit electrons. These emitted electrons transform into photons, enabling us to see the polar lights. Remember: whether it's a rock, a cloud, or a microscopic organism, it doesn't matter—if the light reaching our eyes comes directly from a source rather than being reflected, there is a negative pressure zone at that source. Recall the hand-and-fly analogy! When you loosen your fist, you hear the fly buzzing. Similarly, when dark matter releases an element under negative pressure, you perceive the electrons emitted from that element as heat or light.

97. NATURAL MAGNETS

Natural magnets are naturally occurring minerals with magnetic properties. The most common example is magnetite, a type of iron oxide (Fe_3O_4). It is believed that these minerals have been magnetized by Earth's magnetic field and possess permanent magnetic properties. Natural magnets have been known since ancient times and were used for navigation. The earliest compasses were made using natural magnets. Today, natural magnets play a significant role in geology and magnetism studies. They also provide valuable insights into the nature of magnetic fields and the magnetic structure of our planet.

However, the paragraph above reflects the definition in mainstream science and, according to the Theory of Everything, contains inaccuracies. No natural magnet has been magnetized by Earth's magnetic field. Natural magnets are magnetic due to their symmetrical element structure.

According to the Theory of Everything, a natural phenomenon that should be recognized as a law of physics on Earth is as follows: In a cluster within a medium vibrating at a constant frequency, if symmetrical-shaped elements are in the majority, they exclude the fewer asymmetrical shapes; if asymmetrical elements are in the majority, they exclude the fewer symmetrical shapes. This physical law in the Theory of Everything explains countless phenomena, from the formation of our organs to the formation of natural magnets, providing ample evidence.

Natural magnets are formed due to this law. The reason for the formation of Earth's poles lies in the inside-out positive pressure created by photons from the Sun on Earth. This positive pressure compresses Earth from the north and south, forming the vector lines in Earth's magnetic field.

In natural magnets, the arrangement of elements might be as follows: Symmetrical rectangular prism-shaped elements are organized in cross-molecular alignments among dark matter. This occurs during the clustering of symmetrical rectangular prism-shaped elements in a high-temperature liquid state (such as during lava flow). When the lava cools or the frequency of vibration ceases, the symmetrical elements twist or bend all the dark matter within the cluster simultaneously and in the same direction. This bending of dark matter affects the positions of dark matter in the magnetic field and the magnetic field cage wall.

To explain more clearly: The elements inside a natural magnet are rectangular prism-shaped and symmetrical. Due to their symmetrical structure, as they cool, they influence the spherical dark matter, causing it to take on a slightly rectangular prism shape as well. Consequently, the magnet cluster and its magnetic field cage wall align the dark matter in the surrounding magnetic field—horizontally relative to itself at one end and vertically at the other. When two magnets are brought close with their vertically aligned ends facing each other, they repel; in other positions, they attract. The vectors defined in mainstream science are correct, but their cause is explained above.

Summary of the Cause: The elements forming the magnet distort the shape of the dark matter within the cluster, which in turn distorts the shape of the dark matter forming the magnetic field. This shape distortion in the magnetic field's dark matter creates vector forces in the directions defined by mainstream science.

98. VECTORS IN NATURE

In physics, a vector is a physical quantity that has both magnitude and direction. Vectors are used to describe the position, velocity, acceleration, and forces of objects that move or exert force in multiple dimensions.

Properties of Vectors:

- Magnitude: The length or intensity of the vector, which represents the size of the physical quantity it describes.
- Direction: Indicates the orientation of the vector, typically represented by an arrow drawn from the starting point to the endpoint.

Applications of Vectors:

- Forces: The direction and magnitude of forces are expressed using vectors.
- Velocity and Acceleration: The velocity and acceleration of a moving object are represented by vectors.
- Electricity and Magnetism: Concepts such as electric and magnetic fields are defined using vectors.

Vectors are vital in physics and engineering for the accurate and precise analysis of motion, forces, and other physical quantities.

According to the Theory of Everything, vectors play a significant role in understanding the behavioral mechanics of dark matter, though this has not yet been recognized in mainstream science. While developing this theory, I was unaware of the vector concept used in mainstream physics. Since I proceeded with an inductive approach, learning that the mathematical behavior of dark matter, which emerged during the formation of space, already existed made me happy. So far, and even now, vectors approximately depict the direction and force distribution of dark matter. In this theory, it is predicted that vectors drawn after understanding the structure and other properties of dark matter could lead to a technological revolution across all branches of science.

99. CENTRIFUGAL FORCE

The centrifugal phenomenon is an important physics concept that describes the tendency of a rotating object to move away from the center. This force, often observed in rotating systems, is not a real force but a pseudo-force based on the principle of inertia. When an object rotates, a centripetal force keeps it in a circular orbit. However, from the perspective of the rotating object's moving reference frame, it appears as though the object is being pushed outward by a force; this is called centrifugal force.

For example, when a car takes a sharp turn at high speed, the passengers feel as though they are being thrown outward—this is a result of centrifugal force. The magnitude of this force is directly proportional to the object's rotational speed, mass, and distance from the axis of rotation. The centrifugal phenomenon has significant applications in engineering and daily life, playing a critical role in areas such as centrifuge machines, amusement park rides, and vehicle dynamics.

According to the Theory of Everything, the orbiting of a satellite around a planet or planets around a star is not directly related to centrifugal force. In space, satellites, planets, and stars with massive photon cores have a vortex of dark matter determined by the momentum, speed, and direction of their core at the moment of formation. Vortices within vortices exist in every star system, including our Solar System. However, these space vortices are not like the water vortices we observe on Earth. Since space is entirely filled with dark matter, the vortices of stars, planets, and satellites resemble flattened cylinders suppressed by dark energy. Black holes, too, are vortices through which dark energy passes.

Returning to centrifugal force on Earth: Centrifugal force is caused by the pressure of dark matter. As an object moves on Earth or in space, the dark matter in front of it shifts around the object due to its presence. After the object passes, the dark matter returns to its original position. A car's magnetic field protects its occupants to some extent from this effect. When a car takes a sharp turn at high speed, the passengers' sensation of being thrown outward is due to this phenomenon.

Centrifugal force does not apply to planets and satellites in space because they reside within dark matter vortices.

100. OPTICAL ILLUSION

Optical illusions are misperceptions related to how our eyes and brain interpret visual information. These illusions arise when the images our eyes perceive are interpreted differently by our brain.

According to the Theory of Everything, I categorize optical illusions into two types: those caused by external factors and those generated internally by the brain. Many people have experienced external optical illusions. For instance, on a hot summer day while driving, you may have noticed mirages on the asphalt at a certain angle, resembling water puddles or greenery. Even your reflection in a mirror is an optical illusion. This is a physical phenomenon because the mirage you see is also visible to the person next to you. This occurs when a photon cluster traveling as particles gains external soft atoms (elements) in an environment with intense photon bombardment. Let me explain: Suppose the color gray is normally formed by a photon cluster with 100 elements. As this 100-element photon cluster reflects toward our eyes, it collides with another photon, increasing its element count to 120. As a result, the shade of gray lightens, making it appear as though there's a puddle of water on the asphalt. This is an example of an external optical illusion.

Internal optical illusions, on the other hand, stem from the brain's perception mechanisms. For example, the afterimages you see when you look at a bright light and then suddenly shift your gaze to darkness, or the shifting focus between different shapes when looking at a picture, fall into this category. You can find many images and photos of such illusions online.

The main reason I included this hypothesis is its relevance to astronomy. In astronomy, perhaps half of the distant objects observed might be optical illusions due to mirages and reflections. This possibility is supported by the behavior and structure of photons in the Theory of Everything. A photon is a cluster of soft atoms. As photons travel through dark matter arrays, they lose elements (soft atoms). They typically compensate for these lost elements by absorbing soft atoms from another cluster traveling approximately parallel to them. However, as the distance increases, the rate at which a photon can compensate for its lost elements decreases. This means the photon loses the visual information it carries. Consequently, many of the stars and planets observed in space might be optical illusions, mirages, or reflections.

For example, the North Star (Polaris) is a reflection of the Sun. If the North Star does not have a counterpart in the southern hemisphere, it might be because the galaxy's expansion direction is toward the south. Another example involves the speeds of stars near black holes. When observing stars at the center of the Milky Way, you might see both the star itself and its mirror image reflected by the black hole, leading to perceived anomalies. When viewing a galaxy from above or below, stars near a black hole might appear to be rotating very quickly. However, this could be due to the black hole's vortex distorting and scattering the star's photons. Such illusions might lead to interpretations that the orbital speeds of planets around the Sun in our Solar System differ from the speeds of stars around black holes in the galaxy. This hypothesis might unsettle astrophysicists, but these propositions are made possible by the Theory of Everything.

BOOK FINALE

I'd like to extend a special thank you to everyone who has read this book from start to finish and made it this far. For me, it's far more important that this book is read rather than sold. I have complete confidence in my theory, but I'm less confident in the workings of the scientific community, both in our country and globally. They might ignore this theory or attempt to manipulate it to fit existing frameworks. That's why, if you don't plan to archive this book for your children or yourself, I'd encourage you to gift it to a friend who might read it.

I'm not a professional physicist like many of you might be. Nor do I think I'm more intelligent than you. However, I might have a slightly better imagination and ability to harness that imagination. I'd be delighted if this book brings positive changes to your work and your perspective on nature, contributing to your life in a meaningful way. At the moment, I don't plan to write another book on this theory, but I'd be happy to provide guidance to those who wish to write about it in an academic style. Additionally, I'll do my best to assist readers with any questions they might have about the book via my account on X (formerly known as Twitter) at @mkilic76 . I'd appreciate it if you could share all your feedback—positive or negative—on your social media and discuss the book with other readers.

The greatest strength of my theory is its systematic nature—its coherence as a framework. A framework is much broader than a theory. Once you learn the framework in this theory and integrate it with your own knowledge, you'll have the potential to generate millions of new insights.

Despite my repeated checks before the book's publication, there will undoubtedly be errors that you might notice, and I apologize for those in advance.

Moving forward, I plan to write more books—my next one will be on economics and constitutions, while the others will be works of fiction. Whatever time may bring, I wish you all the best until we meet again—or perhaps not! Take care of yourselves!

Don't miss out!

Visit the website below and you can sign up to receive emails whenever MUSTAFA KILIÇ publishes a new book. There's no charge and no obligation.

https://books2read.com/r/B-A-LZNAE-LKJOG

BOOKS 2 READ

Connecting independent readers to independent writers.

www.ingramcontent.com/pod-product-compliance
Lightning Source LLC
LaVergne TN
LVHW090306210926
848480LV00001B/154